About Island Press

Island Press is the only nonprofit organization in the United States whose principal purpose is the publication of books on environmental issues and natural resource management. We provide solutions-oriented information to professionals, public officials, business and community leaders, and concerned citizens who are shaping responses to environmental problems.

In 2000, Island Press celebrates its sixteenth anniversary as the leading provider of timely and practical books that take a multidisciplinary approach to critical environmental concerns. Our growing list of titles reflects our commitment to bringing the best of an expanding body of literature to the environmental community throughout North America and the world.

Support for Island Press is provided by The Jenifer Altman Foundation, The Bullitt Foundation, The Mary Flagler Cary Charitable Trust, The Nathan Cummings Foundation, The Geraldine R. Dodge Foundation, The Charles Engelhard Foundation, The Ford Foundation, The Vira I. Heinz Endowment, The W. Alton Jones Foundation, The John D. and Catherine T. MacArthur Foundation, The Andrew W. Mellon Foundation, The Charles Stewart Mott Foundation, The Curtis and Edith Munson Foundation, The National Fish and Wildlife Foundation, The National Science Foundation, The New-Land Foundation, The David and Lucile Packard Foundation, The Pew Charitable Trusts, The Surdna Foundation, The Winslow Foundation, and individual donors.

About the American Association for the Advancement of Science

Since its founding in 1848, the American Association for the Advancement of Science (AAAS) has continually worked to advance science. From its early, specific aims concerned with communications and cooperation among scientists, the association's goals now encompass the broader purposes of ". . . furthering the work of scientists, facilitating cooperation among them, fostering scientific freedom and responsibility, improving the effectiveness of science in the advancement of human welfare, advancing education in science, and increasing the public understanding and appreciation of the importance of the methods of science in human progress."

AAAS enrolls more than 145,000 scientists, engineers, science educators, policymakers, and others interested in science and technology who live in the United States and in many other countries throughout the world. In addition, AAAS is the world's largest federation of scientific engineering societies, with 280 organizations that cooperate with the association on a variety of projects, including annual meeting symposia, fellowships, international programs, annual analyses of the federal research and development budget, equal opportunity activities, and science education.

About the Boston Theological Institute

The Boston Theological Institute (BTI), an association of nine theological schools in the greater Boston area, is one of the oldest and largest theological consortia in the United States. It is the only one to include, as constitutive members, schools representing the full range of Christian churches and confessions, as well as representatives of other faiths in its universities. Member schools include Andover Newton Theological School, Boston College Department of Theology, Boston University School of Theology, Episcopal Divinity School, Gordon-Conwell Theological Seminary, Harvard Divinity School, Holy Cross Greek Orthodox School of Theology, Saint John's Seminary, and Weston Jesuit School of Theology. As a consortium, the BTI offers certificates in international mission and ecumenism and science and religion.

CONSUMPTION, POPULATION, AND SUSTAINABILITY

CONSUMPTION, POPULATION, — AND — SUSTAINABILITY

Perspectives from Science and Religion

Edited by Audrey R. Chapman,
Rodney L. Petersen, and Barbara Smith-Moran

ISLAND PRESS
Washington, D.C. • Covelo, California

Another version of "Perspectives on Sustainability" will appear in *Christianity and Ecology: Wholeness, Respect, Justice, Sustainability,* eds. Dieter T. Hessel and Rosemary Radford Ruether (Cambridge: Harvard University Center for the Study of World Religions, in press).

Library of Congress Cataloging-in-Publication Data
Consumption, population, and sustainability : perspectives from
  science and religion / edited by Audrey R. Chapman, Rodney L.
  Petersen, Barbara Smith-Moran.
        p.   cm.
    Includes bibliographical references and index.
    ISBN 1–55963–747–1 (cloth: alk. paper). — ISBN 1–55963–748–X
(pbk. : alk. paper)
    1. Consumption (Economics)—Religious aspects. 2. Consumption
(Economics)—Moral and ethical aspects. 3. Population policy—
Religious aspects. 4. Population policy—Moral and ethical
aspects. 5. Sustainable development—Moral and ethical aspects.
I. Chapman, Audrey R.   II. Petersen, Rodney Lawrence.   III. Smith-
Moran, Barbara.
 HC79.C6C675   2000
 291.1'78366—dc21   99–33887
                                                         CIP

Contents

Preface

The question of global sustainability is one that has begun to absorb the attention and efforts of a number of different communities of scholars and policy personnel over the last quarter century. Consumption levels and issues of population policy directly affect the sustainability equation. Yet both are fraught with deep public policy implications that draw upon two communities that are at times seen to be at odds with one another, the "religious" and the "science" communities, each with its own conversations and social activists.

This volume—and the larger conference and work of which it is a part—attempts to draw these groups into conversation with one another. Therefore, it is appropriate that the book be developed by a scientific organization of such stature as the American Association for the Advancement of Science (AAAS) in collaboration with the Boston Theological Institute (BTI), the association of university divinity schools, schools of theology, and seminaries in the Greater Boston area. Since the inception of this volume, an increasing number of models have been put forward to promote a science–religion dialogue. Much of the credit for this goes to the AAAS and to conversations and studies people associated with AAAS have stimulated. However, if conversation is to proceed, it is clear that there needs to be an adequate understanding of the fundamentals of modern science and its methodology. This must be matched by an informed ethics developed in relation to the contemporary sciences, as well as an understanding of specific issues under consideration. Finally, new methodologies and approaches are needed to delineate proper integration. The authors of this volume argue for the importance of a broad interdisciplinary approach that draws together the concepts raised in the title, which are consumption, population, and sustainability. Short-term environmental policy fails to take into account the economic and moral burdens being placed on future generations through the present depletion of resources.

The manner by which religion, or values independent from the natural sciences, might be integrated is complex. From one point of view, it requires a worldview that is not that of scientism, although it may be scientific. The field of religious studies is one that is deeply divided between those for whom religion is an area of human engagement subject to natural and mechanistic explanation and others for whom such an unreservedly naturalistic approach begs the very premises of religious understanding. Division in the field of anthropology between those who wish to maintain the aloofness of the social sciences from the natural sciences offers insight here. Although the enormity of these debates is only touched on in this book the subject needs to be raised at the outset.

The point raised at mid-century by Nels F. S. Ferré—that the sciences have inadvertently contributed to a collapse of values by fostering a spirit of negativity, detachment, and tentativeness—applies in a special way to the kinds of engaged activity required for environmental sustainability in the twenty-first century.[1] This critique, applied specifically to the emergence of scientific positivism, is one that is challenged today by scholars and policy personnel concerned about and committed to environmental justice and health. A similar concern with respect to technology was developed by the sociologist Jacques Ellul: "Technique is the translation into action of man's concern to master things by means of reason, to account for what is subconscious, make quantitative what is qualitative, make clear and precise the outlines of nature, take hold of chaos and put order into it."[2] This mentality has become, for Ellul, the reigning mythology of our epoch. Finally, Paul Ehrlich offers the opinion that "a quasi-religious movement, one concerned with the need to change the values that now govern much of human activity, is essential to the persistence of our civilization."[3]

Whether Ferré's concern for the sciences, Ellul's focus on technology, or Ehrlich's apprehension toward our environmental future, all three score negatively in the creation of a value-detached objectivity. Together, their concerns promote a science–religion dialogue with the aim of an informed and engaged morality.

Many people have contributed to this book. Thanks goes first to the Aspen Institute and the Pew Charitable Trusts for a grant given to enable a working conference to occur, jointly sponsored by the American Association for the Advancement of Science and the Boston Theological Institute. The conference, entitled "Consumption, Population, and the Environment: Religion and Science Envision Equity for an

Altered Creation," was held November 9–11, 1995. In addition to this volume, the conference facilitated the development of a video, "Living in Nature: Religion and Science in Dialogue on the Environment," a course taught at Andover Newton Theological School, and numerous eddies of conversation and study both in the Boston area and beyond.

In addition to the work of those cited as editors of and contributors to this volume, a special word of thanks goes to Adam Kissel and Forrest Clingerman, successive office operations managers of the Boston Theological Institute, and to Aaron Goldberg from the office of the Program for Science and Religion of the American Association for the Advancement of Science. Rev. Barbara Smith-Moran, E. Kay Bergersen, and Susan Youmans worked extensively with Dr. John Michalczyk, chair of the Fine Arts Department and professor of film studies of Boston College, to produce the video. Colleagues and students associated with the Religion and Ecology Program of the schools of the Boston Theological Institute helped with the conference and, at various stages along the way, with programming that helped to make ongoing work from the conference possible. Special mention goes to William Jones of Andover Newton Theological School; Jonna Higgins, Amy Langston, and Dennis Hargiss of Harvard Divinity School; Amelia Smith of Episcopal Divinity School; and Alexander Kern of Andover Newton Theological School for proofreading assistance. Finally, thanks goes to Dr. Timothy Weiskel, director of the Harvard Seminar on Environmental Values at the Harvard Divinity School. Dr. Weiskel's continued efforts throughout the Harvard University community have helped to keep alive a topic of such philosophical, religious, and ethical import as that of environmental studies.

—Rodney L. Petersen

Notes

1. "The Immortality of Science," *Religion in Life* 10, no. 1 (Winter 1941): 31–40.
2. See his study, *The Technological Society* (New York: Alfred A. Knopf, 1964), p. 43.
3. *The Machinery of Nature* (New York: Simon & Schuster, 1986), p. 17.

Part I
INTRODUCTION

Science, Religion, and the Environment

Audrey R. Chapman*

The creation narratives in the book of Genesis provide images of an ideal landscape that anticipates a balanced and interdependent ecosystem. According to the biblical writers, God brings forth living creatures of every kind, blessing them and telling them to be fruitful and multiply. At the apex of the creation process, the first man and woman are placed in the Garden of Eden to till and keep it. The primordial couple are given one restriction, not to eat of the tree of knowledge of good and evil, but of course they violate this mandate. When they do, they suffer the penalty of banishment. They are driven out of the garden to face the realities of the outside world.

Suppose that the narrative turned out differently, that Adam and Eve were forgiven and allowed to remain in the garden. What might the outcome have been? Beginning the task of procreation, the first couple would soon have had a family. Their children and the countless generations following them would likely have taken God's blessing to be fruitful and multiply quite literally, as would the other species. This process of fruitful procreation would have set up a competition between the human family and otherkind for resources and space. Humans, having eaten of the tree of knowledge, might have tried to resolve the ensuing problems through the ingenious introduction of technology. Over the long term, however, their technology is likely to have been intrusive and affected the ecological balance. Eventually, the

*Audrey R. Chapman is director of the Program of Dialogue on Science, Ethics, and Religion and the Science and Human Rights Program at the American Association for the Advancement of Science in Washington, D.C.

entire environmental system in the garden probably would have been threatened. Instead of banishment to the realities of an often harsh world outside the garden, humans would have undermined their paradise and turned it into a polluted and depleted landscape.

And that is our current situation. The growth in human population from five million people at the dawn of the agricultural revolution to six billion people at the close of the twentieth century has transformed ecosystems and contributed to an environmental crisis that threatens the planet. As ethicist James Nash comments, the biblical injunction to "increase and multiply" may be the only one that humankind has fully obeyed.[1] All other things being equal, a significantly smaller human population, five million people or even one billion, would have had considerably less impact on the environment than the current toll of six billion. If the planet's population had stabilized at even half of our current numbers, it is unlikely that there would be falling water tables, deforestation, or the extinction of thousands of species each year.[2] But of course, not all things are equal. Not only has the human population grown exponentially, but also changing lifestyles and rising consumption patterns have resulted in ever more environmentally damaging technologies. Modern technologies that have been developed to support affluent consumption patterns—even more than increasing numbers—have resulted in such environmental hazards as ozone holes and the possibility of human-induced climate change. This finite planet is being seriously taxed by modern industrial technologies that extract resources and pollute the earth at rates that would have been unimaginable in earlier periods.

What are the implications? As theologian Daniel Maguire starkly summarizes the situation, "If current trends continue, we will not. And that is qualitatively and epochally new."[3] As the twentieth century is coming to an end, many scientists believe that the cumulative impact of human activity is pushing to the limit the earth's life-supporting or carrying capacity, perhaps even exceeding it. Our Eden is being threatened, but there is no place to go.

This volume is an outgrowth of "Consumption, Population and the Environment: Religion and Science Envision Equity for an Altered Creation" a conference cosponsored by the Boston Theological Institute and the American Association for the Advancement of Science. Held in November 1995 in Weston, Massachusetts, the conference brought together more than 250 scientists and people of religious faith to discuss scientific and religious perspectives regarding the impact of consumption patterns and population trends on the environment; to envision alternative and more equitable value systems, economic arrangements, and technologies; and to consider the potential contributions of religious communities to developing a more sustainable future. The focus of the conference and this volume raise two questions: First, why have a dialogue between the scientific and religious communities relating to the environment, and, second, why should such a dialogue focus on the issues of consumption and population?

Relationship between Science and Religion

The philosopher Alfred North Whitehead once described science and religion as "the two strongest general forces (apart from the mere impulses of the various senses) which influence men."[4] Taking up this theme, a World Council of Churches publication titled *Faith, Science and the Future* characterizes faith and science as two human ventures: "One meets the world with an inquiring intelligence. It values accurate, testable knowledge. It experiences the sheer joy of knowing, of understanding the world, of making discoveries and the power of prediction and control."[5] This, of course, is the venture of science and technology, but, as the authors remind us, in some ways science is also a venture of faith, faith in the ultimate rationality and knowability of the world. Religious faith, in contrast, "meets the world in wonder, trust and commitment. It values the relations of persons to each other and to their ultimate source and destiny. It glories in the beauty of holiness and the responsibility of service."[6] And it might be added that religious faith also pursues truth and understanding—however, with different methodologies and approaches than science does. Although both science and religion are essential to human understanding, their relationship has sometimes been problematic.

The relationship between the two ventures has frequently been portrayed as one of warfare, and various historical instances of conflict are well known. The Vatican's condemnation of Galileo for accepting the Copernican view of a heliocentric (sun-centered) universe is often cited as one example. The decision to burn at the stake Giordano Bruno, a sixteenth-century astronomer and philosopher, because of the tenacity with which he maintained his unorthodox ideas of an infinite universe and a multiplicity of worlds also constitutes part of the litany of the supposed incompatibility of religious and scientific principles. A third example is that of the 1925 Scopes trial, in which it was argued that the teaching of evolution in the schools should be forbidden because it is contrary to scripture.

However, the conflict model simplifies and distorts a complex historical relationship. Historians remind us that modern empirical western science "was cast in a matrix of Christian theology. The dynamism of religious devotion, shaped by the Judeo-Christian dogma of creation, gave it impetus."[7] According to Lynn White, by the early thirteenth century in the Latin West, natural theology was moving away from a focus on decoding the physical symbols of God's communication with humanity and was making more of an effort to understand God's mind by discovering how the creation operates. Up to and including Leibnitz and Galileo, scientists typically explained their motivation in religious terms. Not until the late eighteenth century did scientists secularize their vocations.[8]

Even in the post–Enlightenment period, it can be argued that autonomy rather than conflict was more often the dominant theme in the relationship between science and religion. After all, the dynamic of secularization over the centuries was to push religion from the public to the private sphere and to emphasize the contrast between faith and reason. Many writers in the history of Western thought have dealt with the epistemological dichotomy between religious and scientific knowledge, each of which is said to have its own distinctive domain and methods. The prevailing metaphysical dualism of spirit and matter further reinforced the independence and mutual autonomy of the two fields. Langdon Gilkey's testimony at a 1982 trial provides a contemporary expression of this view. The trial contested an Arkansas law requiring the teaching of "creationist theory," that there is scientific evidence for so-called scriptural claims that the world was created within the last few thousand years, as scientific theory in high school biology classes. Gilkey, who is a theologian, makes the following distinctions: science asks objective *how* questions, while religion asks *why* questions about meaning and purpose in the world and about our origin and ultimate destiny. Science seeks to explain objective, public, repeatable data; religion is concerned with the existence of order and beauty in the world and the experiences of our inner life. Logical coherence and experimental adequacy provide the basis of authority in science; the divine is the final authority for religion, and revelation through God's human agents constitutes the medium of enlightenment and insight. Science makes quantitative predictions that can be tested experimentally; religion uses symbolic and analogical language.[9]

Rather than ongoing warfare, conflicts between science and religion have been sporadic, usually occasioned by scientific discoveries that threaten religious dogma. With the exception of the relatively small group of biblical literalists, most modern religionists, at least in the West, have repented of the famous, or perhaps infamous, efforts to suppress scientific findings. In 1984, a Vatican commission acknowledged that "church officials had erred in condemning Galileo" in the seventeenth century.[10] In 1988, Pope John Paul II issued a statement in which he underscored the importance of the search for "areas of common ground" between the fields of religion and science. According to the pope, "It is crucial that this common search based on critical openness and interchange should not only continue but also grow and deepen in its quality and scope. For the impact [that science and religion have] and will continue to have, on the course of civilization and on the world itself, cannot be overestimated, and there is so much that each can offer the other."[11] While critical of scientism, the philosophical notion that does not admit the validity of forms of knowledge other than those of the sciences, the pope's recent encyclical letter *Fides et Ratio* conveys his admiration for scientific achievements and offers "encouragement to these brave pioneers of scientific research, to whom humanity owes so much of its current development." He urges scientists "to continue their efforts without ever abandoning the sapiential horizon within which scientific and technological achievements are wedded to the philosophical and ethical values which are the distinctive and indelible mark of the human person."[12]

One dimension of an effort at rapprochement is the religious community's growing interest in science. Efforts to better understand the implications for theology of scientific methodologies and discoveries are manifested on an academic level in course offerings, research, and literature addressing various aspects of "theology for a scientific age."[13] Theologians from many different faith traditions have begun to explore the implications of modern science for understanding the nature and purpose of the creation. Several academic programs and centers have been established, one of which, the Faith and Science Exchange at the Boston Theological Institute, played a major role in the cosponsorship of the conference that provided the basis for this volume. Religious ethicists, and in some cases faith communities, have also sought to address ethical and policy issues related to the impact of science on society. On the other side of the divide, scientists in some fields, particularly physics, cosmology, and astronomy, have begun to consider issues related to the origins, nature, and ultimate destiny of the universe, which are traditionally within the religious domain.

These developments raise a question as to how the religious and scientific communities can develop more meaningful and constructive relationships. In his landmark volume, *Religion in an Age of Science*, Ian Barbour identifies four options for the relationship between science and religion: conflict (the "warfare" model), independence (which assumes that the methodologies and subject areas of religion and science are unrelated to one another), dialogue (the effort to explore boundary questions), and integration (approaches to developing a comprehensive metaphysics or coherent worldview based on science and religion).[14] Likewise, John Haught's book *Science & Religion: From Conflict to Conversation* offers four typologies: conflict, contrast, contact, and confirmation, the first three of which are similar to Barbour's options. Haught's fourth category, confirmation, expresses his belief that religion ultimately inspires and facilitates scientific discovery by setting the framework for a rational and orderly universe.[15]

There are several other complex and nuanced typologies. Ted Peters outlines eight models of interaction: (1) scientism (the belief that science offers the only path to knowledge); (2) scientific imperialism (which claims that knowledge of the divine comes from scientific research rather than religious revelation); (3) ecclesiastical authoritarianism (religious censorship of science); (4) scientific creationism (the use of pseudoscience to claim that biblical accounts of creation are fully scientific); (5) the two-language theory (which distinguishes between the language of fact and the language of values); (6) hypothetical consonance (which looks for areas in which there is a correspondence between science and religion); (7) ethical overlap (which addresses challenges created by science and technology); and (8) new age spirituality (which integrates spirituality with physical theory).[16] Philip Hefner's survey of contemporary thinking on the religion–science interface offers six trajectories: (1) the Christian evangelical option (which reaffirms the rationality of traditional belief); (2) the modern option (which translates traditional religious wisdom into scientific concepts); (3) the constructivist traditional option (which interprets science by means of traditional theological concepts that have themselves been reinterpreted in light of scientific findings); (4) the postmodern–new age option (which constructs new science-based myths); (5) the critical post-Enlightenment option (which expresses the truth at the "obscure margin" between what we know through science and the transcendent reality that we seek to know); and (6) the postmodern constructivist option (which fashions a new metaphysical loom on which scientific knowledge can be woven).[17]

Media reports notwithstanding, few mainstream religious thinkers currently believe that there are intrinsic conflicts between contemporary science and classical Western religious teachings. The major and sometimes very vocal exceptions within the religious community are fundamentalists and scriptural literalists who believe that every word and passage in the Bible are literally true and must be considered to be scientifically factual. So-called creation scientists, or creationists, who purport to prove the scientific basis of biblical accounts of the creation and disprove scientific theories of evolution are a subgroup within the "conflict" constituency. Scientific skeptics who assume that modern science disproves religious belief in a purposeful universe created by a benevolent God constitute another subcommunity. Some scientists make the epistemological claim that science and scientific methods offer the only reliable guide to truth. Others go beyond an epistemological scientism to assert that physical matter is the only or fundamental reality in the universe.[18] Barbour and Haught argue that earlier conflicts between the religious and scientific communities, as well as current attitudes of religious fundamentalists and scientific skeptics, result from mistakenly conflating or confusing the appropriate methodologies and subject matter of science and religion.[19]

Recently, some religious thinkers and scientists have sought opportunities for constructive dialogue and mutual enrichment between the two disciplines. This reaching beyond the independence and autonomy of science and religion sometimes reflects the desire to explore boundary questions, such as the relationship between Judeo-Christian religion and the development of modern science.[20] Other times it advocates greater contact, anticipating that scientific knowledge can broaden the horizons of religious faith, while the perspective of religious faith can contribute to a fuller understanding of the universe.[21] Discovery of unanswered questions or issues that lie outside the competence of one's field can serve as a stimulus. Relevant examples are genetic scientists' realization that their research raises ethical dilemmas and fundamental questions as to what it means to be human, and theologians' interest in the implications of contemporary physics and cosmology for understanding the nature and destiny of the universe. There are also instances in which scientists and/or religious leaders, confronting serious problems or issues that affect the future of their country or the planet, determine that political cooperation between the communities can enhance prospects for a satisfactory resolution.

Despite the interests of individual scientists, the scientific community has generally been reticent about cross-disciplinary initiatives and interface between the fields. Assumptions not only that science and religion have different epistemologies and methodologies, but also that the scientific approach has greater value or truth, constitute a major barrier. It should be noted that the willingness of scientists to collaborate with religious leaders, even in some instances in which the scientific community has taken the initiative, does not necessarily mean that scientists respect their views. John Haught, a theologian who was a participant in the meeting that drafted the statement discussed later in this essay entitled "Joint Appeal by Science and Religion on the Environment," notes the anomaly of three prominent scientists, well known for their positions that religions are essentially illusory, appealing to the religious leaders present to foster the moral inspiration needed to protect the environment. Haught questions whether it is intellectually honest for these scientists to attempt to co-opt the moral enthusiasm of religions for the sake of ecology, especially since the source of the inspiration is the very religious symbols and ideas that they consider to be untrue and inappropriate.[22] Under such circumstances it is difficult to have a meaningful dialogue because, at a minimum, successful interaction presumes mutual respect. As David Byers observes, "Unless the scientific community treats religious principles as seriously as their own data, the deck is stacked from the beginning and dialogue is fruitless."[23]

The inability of most religious thinkers to comprehend scientific literature or to communicate easily within a scientific idiom quite understandably discourages interest among scientists. To date, most dialogues and joint activities have required that scientists operate within a religious or ethical field of reference rather than developing a context for research and reflection that is truly multi- or cross-disciplinary. Few theologians and theological ethicists are expert in a field of science, and many are not even scientifically literate. Most of those who are expert in both theology and science are trained scientists who turned to religion as a second career.

Nevertheless, there are signs of growing interest within the scientific community to reach across the historic divide with the religious community. While still a small minority, some seminaries and graduate religion departments have increasing numbers of students with scientific backgrounds, who are taking courses on religion and theology to enrich their own understanding and who are even enrolling in degree programs. The John Templeton Foundation's innovative program to promote the teaching of courses on the interaction between science and religion has stimulated the development of such course offerings at several universities, many of them team-taught by scientists and religion scholars. The establishment of such centers as the Faith and Science Exchange in Boston, the Center for Theology and the Natural Sciences at the Graduate Theological Union in Berkeley, the Institute for Theological Encounter with Science and Technology in St. Louis, and the Chicago Center for Religion and Science has provided fora for meaningful intellectual exploration between the fields and has promoted the development of collegial relationships among their staff and the participants in their programs. Two membership organizations, the Institute for Religion in an Age of Science and the American Scientific Affiliation, sponsor programs and publish journals that encourage broad discussion of and research into the relationship between religion and science.

The decision of the American Association for the Advancement of Science, the largest scientific membership organization in the world, to establish a Program of Dialogue on Science, Ethics, and Religion, which I direct, is another encouraging development. The program has three major objectives: (1) to promote knowledge about developments in science and technology within the religious community; (2) to provide opportunities for dialogue among members of the scientific and religious communities; and (3) to promote collaboration between members of the scientific and religious communities on projects that explore the ethical and religious implications of scientific developments. By offering scientists a vehicle for collaboration based in a scientific institution, the Program of Dialogue on Science, Ethics, and Religion may be able to encourage their participation in cross-disciplinary discussions and multidisciplinary projects.

Meetings, Initiatives, and Dialogues on Environmental Issues

The past thirty years have been marked by increasing awareness of and heightened concern for the impact of human societies on our planet. In the 1960s, a few economists began to assess the hitherto unrecognized costs of economic growth to the environment. Kenneth Boulding, an economist concerned about the perils of the "reckless cowboy economy," appealed to the National Council of Churches to promote an ethic of moderation, conservation, and recycling.[24] In 1967, Lynn White, a medieval historian, published an article in *Science* magazine entitled "The Historical Roots of Our Ecological Crisis," in which he indicted Western Christianity and modern science and technology as two interrelated causes of contemporary environmental problems.[25] Two landmark studies in the 1970s raised the argument that modern industrial society is not sustainable. The first, *A Blueprint for Survival*,[26] was prepared by a group of British scientists and philosophers, and the second, *The Limits to Growth*,[27] was produced at the Massachusetts Institute of Technology under the sponsorship of the Club of Rome, an international group of scientists and industrialists. Based on its finding that five principal factors—population, capital, food, consumption of nonrenewable resources (including energy), and pollution—were growing exponentially at rapid rates, the latter study concludes that if present trends continue unchanged, "the limits to growth on this planet will be reached sometime within the next one hundred years."[28] The United Nations held the first of a series of international conferences on the environment in 1972 in Stockholm. As well as detailing the global character of environmental problems, that conference acknowledged that environmental protection must become an essential element in social and economic development.[29]

As early as the mid-1960s, awareness of looming environmental problems led a group of Christian theologians, scientists, and church leaders to form the Faith-Man-Nature study group under the aegis of the National Council of Churches. Its initial work focused on remedying specific environmental problems and changing religious attitudes toward nature. Later the group determined that the scope of the environmental crisis raised major technological-economic and social-political issues. By the 1970s, an eco-justice movement seeking to integrate ecology, justice, and Christian faith began to find expression in various theological, ethical, biblical, historical, and public policy studies in North America.[30]

Paralleling these activities, in 1966 the World Conference of Churches (WCC) convened a conference entitled "Christians in the Technological and Social Revolutions of Our Time." Four years later, the WCC's Department of Church and Society launched a five-year study program giving particular attention to science and the problems of worldwide technological change, one component of which was the environmental crisis. A week-long conference took place in June 1970 to clarify the issues and propose priorities for the new program; over one hundred participants from six continents were invited. During much of the conference, theologians and church leaders, who constituted a minority of the participants, listened as scientists and planners sketched out the technical and human problems that contemporary societies confront. Environmental issues figured prominently in the discussions as well as in the conference report.[31] The culminating report of the five-year study, adopted at Bucharest in 1974, accepted the thesis of nature's limits and called for a society that is both just and sustainable. Two years later in Nairobi, the Fifth Assembly of the WCC adopted these recommendations and established a council-wide program entitled "Just, Participatory, and Sustainable Society."[32]

In July 1979, the WCC Working Group on Church and Society sponsored the "Conference on Faith, Science and the Future" held at the Massachusetts Institute of Technology. Charles Birch, a biologist at the University of Sydney, Australia, and vice chairman of the WCC's Working Group on Church and Society, played a major role in the conference. His address entitled "Nature, Humanity and God in Ecological Perspective" advocated for a new partnership of faith and science based on an ecological view of nature, humanity, and God. According to Birch, science is a social construct: the dominant worldview derived from science during any period reflects the nature of the science that is practiced. He claimed that contemporary society has a mechanistic, scientific–technological worldview, which views nature and the living organisms within it as a machine. Birch traced this perspective to the legacy of a society bent on mastery over and control of nature. Christian theology, particularly in the West, by accommodating the mechanistic cosmology of science, has detached nature, humanity, and God. To meet the contemporary challenge of the ecological crisis and to provide a more ecologically relevant grounding, Birch proposed that faith and science form a new partnership that acknowledges the oneness of nature, humanity, and God and that supports an ethic of infinite responsibility to all life.[33]

Other faith–science initiatives on the environment have emanated from the scientific community. In January 1990, thirty-four well-known scientists, among them Carl Sagan, Hans Bethe, Freeman Dyson, Stephen Jay Gould, Henry Kendall, and Jerome Wiesner, issued "An Open Letter to the Religious Community" to encourage a spirit of common cause and joint action to preserve the earth.[34] Like other reviews of the state of the environment, the letter details examples of environmental degradation and alteration: depletion of the protective ozone layer, global warming unprecedented in the last 150 millennia, the obliteration of an acre of forest every second, the rapid extinction of species. While acknowledging that the activities that gave rise to these dangers may have been well meaning, the letter warns that they now imperil our species and many others. According to the letter, these transnational, transgenerational, and transideological assaults on the environment constitute what in religious language is sometimes called "crimes against creation."

"An Open Letter to the Religious Community" emphasizes that both the religious and the scientific communities have vital roles to play in saving the environment. It identifies two rationales for the religious community's involvement. First, the environmental crisis requires radical changes, and many of these far-reaching, long-term, and potentially effective approaches to protect the environment are likely to encounter widespread inertia, denial, and resistance. As with peace, human rights, and social justice issues, religious institutions can be a strong force in encouraging national and international efforts to help preserve the earth. Second, as stated in the letter, "Efforts to safeguard and cherish the environment need to be infused with a vision of the sacred." In one of its most moving passages, the letter states, "As scientists, many of us have had profound experiences of awe and reverence before the universe. We understand that what is regarded as sacred is more likely to be treated with care and respect. Our planetary home should be so regarded." The letter goes on to note that if the environment is to be safeguarded and cherished, such efforts will require a much wider and deeper understanding of science and technology. "If we do not understand the problem, it is unlikely we will be able to fix it."[35]

Dissemination of the letter coincided with accelerated activity by many American faith communions and denominations on the environment. During the 1980s and early 1990s, many national religious bodies adopted policy statements identifying the environmental crisis as a religious issue. One of the best known of these, the U.S. Catholic bishops' pastoral statement "Renewing the Earth: An Invitation to Reflection and Action on Environment in Light of Catholic Social Teaching," was issued in November 1991. Also in 1991, the National Council of Churches established an office on environmental and economic justice; the Episcopal Church funded its first program on environment and sustainable development; World Vision, an evangelical Christian development and relief organization, hosted an environmental conference; and the United Church of Christ organized an environmental summit for people of color. As testimony to the "greening" of the religious community, in May 1992 the U.S. Catholic Conference, major communions within the National Council of Churches, the Consultation on the Environment and Jewish Life, and the Evangelical Environmental Network formed the National Religious Partnership for the Environment. The goal of the partnership is to broaden exponentially the base of mainstream commitment, integrate issues of social justice and environmental protection, and urge behavioral change in the lives of congregants.[36]

In addition, there has been a remarkable outpouring of religious writing on environmental issues. Religiously grounded theologians and ethicists—Christian, Jewish, and Muslim—have sought to respond to the current crisis and in some cases to reorient and reformulate traditions to make them more consistent with ecological thinking. A review of Protestant thought on the environment from 1970 to 1990 could be applied more broadly across the religious spectrum. According to Robert Booth Fowler, there has been a major shift during the past several decades toward environmental concerns within Protestantism. He concludes that there is a considerable consensus on the necessity of action by Christians to address environmental issues, but that there are disagreements over how to conceptualize and act on a Christian ecological agenda.[37] A critical literature survey of the period 1961–93, entitled *Ecology, Justice, and Christian Faith*, similarly observes that "the struggle to integrate ecology, justice, and Christian faith that began in earnest in the 1960s has every indication of being a permanent part of Christian thought and practice for many years to come."[38]

This growing consciousness within the religious community about the urgency of responding to the environmental crisis prompted a group of religious leaders, among them Joseph Cardinal Bernardin, Archbishop Iakovos, Robert Shuller, and Elie Wiesel, to welcome the 1990 letter of invitation from the scientific community as "a unique moment and opportunity in the relationship of science and religion."[39] During the next two years, several efforts were made to facilitate consultations between religious leaders and scientists in meetings organized under the title "Joint Appeal of Science and Religion for the Environment." These meetings led up to a major conference in Washington in May 1992 of more than 150 religious leaders and scientists. The conference combined briefings by eminent scientists with a Senate committee hearing and a press conference organized by then Senator Albert Gore. A committee at the conference (in which this author participated) drafted a statement entitled "Declaration of the Mission to Washington of the Joint Appeal by Religion and Science for the Environment," which was signed by most of the participants at the conference.

The text of the statement refers to the history of conflict between the two communities but emphasizes the need to overcome differences in order to address environmental problems. It identifies the signatories to the statement as "people of faith and science who, for centuries, often have traveled different roads." It goes on to note, "In a time of environmental crisis, we find these roads converging. As this meeting symbolizes, our two ancient, sometimes antagonistic, traditions now reach out to one another in a common endeavor to preserve the home we share." Like the earlier open letter to the religious community, the text enumerates a litany of environmental abuses: tampering with the climate; thinning the ozone layer and creating holes in it; poisoning the air, the land, and the water; destroying forests, grasslands, and other ecosystems; and causing the extinction of species at a pace not seen since the end of the age of the dinosaurs. Affirming the possibility of, indeed the need for, collaboration, the signatories state that they "believe that science and religion, working together, have an essential contribution to make toward any significant mitigation and resolution of the world environmental crisis."[40] This is so because:

> What good are the most fervent moral imperatives if we do not understand the dangers and how to avoid them? What good is all the data in the world without a steadfast moral compass? . . . Insofar as our peril arises from a neglect of moral values, human pride,

arrogance, inattention, greed, improvidence, and a penchant for the short term over the long, religion has an essential role to play. Insofar as our peril arises from our ignorance of the intricate interconnectedness of nature, science has an essential role to play.[41]

Despite this rhetoric, there have been relatively few subsequent joint initiatives of the two communities on the environment with concrete products. The National Religious Partnership for the Environment opened a science office headquartered at the Union of Concerned Scientists, but the role of the office was more to serve as a scientific resource to religious communities than to sponsor joint projects or organize ongoing dialogues. Environmental groups such as the World Wildlife Fund have sponsored interreligious meetings, but such conferences have not led to ongoing activities. The culminating conference in a 1997–98 series of conferences on religions of the world and ecology held at Harvard University did include several prominent scientists. Calls for interdisciplinary dialogue in the writings of individual religious ethicists and theologians, however, frequently remain just that. Forums that have brought together scientists and people of religious faith to discuss the environment generally have affirmed a common interest in preserving the environment and improving mutual understanding without spelling out any specifics.

When the staffs of the Boston Theological Institute and the American Association for the Advancement of Science considered organizing a conference that would bring together scientists and religious thinkers on the environment, it was decided that one way to encourage more meaningful dialogue was to have a sharper focus on the subject matter—that is, to deal in some depth with one dimension of the environmental crisis rather than the whole spectrum of issues. It was felt that greater focus would enable the two communities to reflect and communicate in ways that were mutually comprehensible and enriching. By narrowing the scope of the dialogue, the scientific community would better be able to consider the ethical, policy, and religious significance of relevant scientific findings and the religious community would find it more feasible to engage in scientifically informed theological and ethical analysis. Once this strategy was decided on, the next issue was to determine the topic.

Consumption, Population, and Sustainability

The topic of the interrelationships of population pressure, consumption patterns, and environmental sustainability constitutes a particularly appropriate focus for a science–religion dialogue both because of its significance and because the subject requires the combined knowledge and insights of the two communities. Behind much of the flowering of work on the environment by religious and secular thinkers is the realization that humanity threatens not only the web of life on Earth but also its own survival. According to these assessments, "We are exceeding the biosphere's carrying capacity—i.e., we are overloading the planet's 'metabolic' capacity to absorb, replenish, and restore."[42] The growth in human population, affluent consumption patterns, and associated economic activity is placing great stress on the earth's environmental support systems. Expanding human demands on the environment have led to land degradation, severe pollution damage, loss of biological diversity, and extensive deforestation. Other signs that nature's limits are beginning to impose themselves on the human agenda include falling water tables, resulting in water scarcity in many areas; declining bird and wildlife populations; collapsing fisheries; record heat waves; and reductions in grain harvests. Some forms of severe environmental damage, such as the production of greenhouse gases at rates too high to be absorbed by the world's oceans, are likely to become more serious problems in the years ahead.[43]

Confronted with the peril of large-scale ecological imbalances, humanity faces a series of crucial choices in order to survive. Ethicist Roger Shinn describes such decisions as "forced options." His terminology reflects the need to meet these urgent demands immediately through defining the issues, clarifying the decision-making processes, and proposing specific solutions. To postpone doing so, according to Shinn, is to make a decision by default, often with grave and unforeseen repercussions. While he acknowledges that societies in the past, knowingly or not, also faced forced options, current technological powers increase the pace of events and the scope of their impact. The fact that the risks are more global in scale and more immediate in consequence changes the nature of ethical responsibility. It also precludes using the method of trial and error because failure could be catastrophic.[44]

How do we individually and collectively approach such momentous decisions? The complexity and significance of the forced options we confront become even more daunting in the absence of defined methodologies and social mechanisms for deliberation and decision making. In many instances, the future will be shaped more by the collective import of many individual, family, or group decisions than by national or global determinations. In making these decisions, Shinn underscores the need to combine ethical and technical judgments and to meld moral responsibility and scientific knowledge.[45] He emphasizes that ecological sustainability, global justice, and the human future are profoundly religious issues. They touch upon some of the deepest human matters of our relationship to one another and to the nature from which we spring and of which we are a part.[46] However, as Shinn recognizes, religion makes and exacerbates these issues as often as it contributes to their resolution. And there are no easy answers from religion. According to him, "There is no help in merely mouthing old religious answers to questions that the past never faced in their present form and scope—in some cases questions that the past never faced at all."[47]

The potential contributions of science and religion, as stimulated through a dialogue process, to understanding and resolving these forced options are far greater than the simple sum of their work done in isolation from one another. The findings of scientific research are central to comprehending the impact of human populations on the environment and the implications for the welfare of current and future generations, but, left to determine their own agenda, scientists may not focus on such policy-relevant issues. Even if they do, decision makers may not act on their findings, and members of society may lack motivation to undertake the required behavioral changes. A more ecologically sustainable future requires a willingness to address these problems by radically altering lifestyles, refashioning technologies, and, in affluent countries, adopting more frugal consumption patterns. This needed revolution in human values in order to sustain the creation, bring about greater equity, and protect the interests of future generations falls potentially within the domain of the religious community. However, it is not clear whether the leaders of contemporary faith communions acknowledge the gravity of the situation or, if they do, have the ability to foster such a revolution. Their members seem even less inclined to accept circumscribed lifestyles for the sake of those who otherwise will go without even the basic needs of life.

Organization of This Volume

For the reasons noted previously, it is critical that societies better understand and respond to the impact of population growth and consumption patterns on environmental sustainability. The scientific and religious communities can both make important contributions to this process. And, as mentioned, the two communities can enhance their contributions to the creation of a more equitable future through dialogue and collaboration.

The essays in this volume come from a variety of sources. A majority are revised versions of texts given at the conference; in a few cases, conference papers were first published elsewhere and are reprinted here. To provide balance and coverage of important topics, particularly in the scientific field, essays were drawn from additional sources, one of which was the annual meetings of the American Association for the Advancement of Science and other conferences dealing with issues related to population, consumption, and environmental sustainability. Several essays were commissioned for the volume from other scholars.

As in the case of most edited volumes, the essays reflect a diversity of perspectives, not all of which are consistent with one another. There is also some overlap and repetition in the subject matter covered.

The volume is organized into five parts: (1) introduction, (2) scientific perspective, (3) religious and theological perspectives, (4) ethics and public policy issues, and (5) conclusion. Parts 2 through 4—science, religion and theology, and ethics and public policy—include an introduction that provides an overview of the issues covered in the section. Since parts 1 and 5 each consist of only two essays, it did not seem necessary to provide an overview of their contents.

Notes

1. James A. Nash, *Loving Nature: Ecological Integrity and Christian Responsibility* (Nashville: Abingdon Press, 1991), p. 44.
2. This analysis from the World Resources Institute's, *1994 Environmental Almanac* was cited in World Wildlife Fund, *Population and the Environment* (Gland, Switzerland: WWF, 1994), p. 9.
3. Daniel C. Maguire, *The Moral Core of Judaism and Christianity: Reclaiming the Revolution* (Minneapolis: Fortress Press, 1993), p. 13.
4. Quoted in Paul Abrecht, ed., on behalf of the World Council of Churches, *Faith, Science and the Future* (Philadelphia: Fortress Press, 1978), p. 11.
5. Ibid., p.11.
6. Ibid., p.11.
7. Lynn White Jr., "The Historical Roots of Our Ecological Crisis," *Science* 155 (March 10, 1967): 1205.
8. Ibid.
9. Langdon Gilkey, *Creationism on Trial* (Minneapolis: Winston Press, 1985), pp. 108–16, cited in Ian Barbour, *Religion in an Age of Science, The Gifford Lectures, 1989–1991*, vol. 1 (San Francisco: HarperSanFrancisco, 1990), p. 12.
10. Quoted in Barbour, *Religion in an Age of Science*, p. 8.
11. Quoted in David M. Byers, "Religion and Science: The Emerging Dialogue," *America* (April 20, 1996), p. 9.
12. Encyclical Letter *Fides et Ratio* of the Supreme Pontiff John Paul II to the

Bishops of the Catholic Church on the Relationship between Faith and Reason, par. 106.

13. *Theology for a Scientific Age* is the title of a seminal work by Arthur Peacocke based in part on the author's 1993 Gifford Lectures (Minneapolis: Fortress Press, 1993).

14. Barbour, *Religion in an Age of Science,* pp. 3–30.

15. John F. Haught, *Science & Religion: From Conflict to Conversation* (Mahwah, N.J.: Paulist Press, 1995).

16. Ted Peters, "Theology and Science: Where Are We?" *Zygon: Journal of Religion & Science* 31 (June 1996): 323–43.

17. Philip Hefner, "Science and Religion and the Search for Meaning," *Zygon: Journal of Religion & Science* 31 (June 1996): 307–322.

18. Ibid., pp. 15–16.

19. Barbour, *Religion in an Age of Science*, pp. 4–10, and Haught, *Science and Religion*, pp. 14–19.

20. Barbour, *Religion in an Age of Science*, p. 17.

21. Haught, *Science and Religion*, p. 18.

22. John F. Haught, *The Promise of Nature: Ecology and Cosmic* (Mahwah, N.J.: Paulist Press, 1993), pp. 8–10.

23. Byers, "Religion and Science," p. 10.

24. Kenneth E. Boulding, "The Wisdom of Man and the Wisdom of God," in *Human Values on the Spaceship Earth* (New York: National Council of Churches, 1966), cited in Roger Lincoln Shinn, *Forced Options: Social Decisions for the 21st Century* (San Francisco: Harper & Row, 1982), p. 111.

25. White, "The Historical Roots of Our Ecological Crisis," 1203–07.

26. Edward Goldsmith et al., eds., *A Blueprint for Survival* (Boston: Houghton Mifflin, 1972).

27. Donella H. Meadows, Dennis L. Meadows, Jorgen Randers, and William W. Behrens, III, *The Limits to Growth* (New York: Universe, 1972).

28. Ibid., p. 24.

29. Peter W. Bakken, Joan Gibb Engel, and J. Ronald Engel, "A Critical Survey," in *Ecology, Justice, and Christian Faith: A Critical Guide to the Literature* (Westport, Conn.: Greenwood Press, 1995), p. 8.

30. Ibid., pp. 6–8.

31. David M. Gill, *Technology, Faith, and the Future of Man: From Here to Where?* (Geneva: World Council of Churches, 1970).

32. Bakken, Engel, and Engel, "A Critical Survey," p. 9.

33. Charles Birch, "Nature, Humanity and God in Ecological Perspective," in Roger L. Shinn, ed., *Faith and Science in an Unjust World: Report of the World Council of Churches' Conference on Faith, Science and the Future*, vol. 1: Plenary Presentations (Geneva: World Council of Churches, 1980). pp. 62–73.

34. "An Open Letter to the Religious Community," January 1990, available from the Science Office of the National Religious Partnership for the Environment, P.O. Box 9105, Cambridge, MA 02238.

35. Ibid.

36. National Religious Partnership for the Environment, "History and Organizational Background." The National Religious Partnership for the Environment is headquartered at the Cathedral of St. John the Divine, 1047 Amsterdam Avenue, New York, NY 10025 (800-435-9466).

37. Robert Booth Fowler, *The Greening of Protestant Thought* (Chapel Hill: University of North Carolina Press, 1995), p. 175.

38. Peter W. Bakken, Joan Gibb Engel, and J. Ronald Engel, *Ecology, Justice, and Christian Faith: A Critical Guide to the Literature* (Westport, Conn.: Greenwood Press, 1995), p. 4.

39. Ibid.

40. "Declaration of the Mission to Washington," Joint Appeal by Religion and Science for the Environment, available from the National Religious Partnership for the Environment's Science Office, P.O. Box 9105, Cambridge, MA. 02238. It is also reprinted in Roger S. Gottlieb, ed., *This Sacred Earth: Religion, Nature, Environment* (New York: Routledge, 1996), pp. 640–42.

41. Ibid.

42. A.J. McMichael, *Planetary Overload: Global Environmental Change and the Health of the Human Species* (Cambridge, England: Cambridge University Press, 1993), p. 1.

43. Lester R. Brown, "Nature's Limits," Worldwatch Institute, *State of the World 1995* (New York: W.W. Norton, 1995), pp. 3–20.

44. Roger Lincoln Shinn, *Forced Options: Social Decisions for the 21st Century* (San Francisco: Harper & Row, 1982), pp. 1–11.

45. Ibid., p.11.

46. Ibid., p. 184.

47. Ibid.

Perspectives on Sustainability

Ian G. Barbour*

Members of the scientific and religious communities have been working together in recent years in a common concern for the environment and the future of life on planet Earth.

Such collaboration is in part the result of pragmatic considerations: People who share a concern can be more effective in education and political action if they work together. But I believe there are also deeper reasons the scientific and religious communities should work together on issues of population, consumption, and the environment. Both communities have a significant contribution to make to their common endeavor; moreover, interdisciplinary dialogue is motivated by recognition that the problems of the real world cut across the boundaries of academic disciplines. Nevertheless, I believe it is important to note the characteristic perspective of each discipline in order to clarify what we are doing and how we can learn from each other. There are also many convictions that we share, even though they derive from differing sources.

In this essay, I will consider contributions primarily from science, contributions jointly from science and religion, and finally some distinctive contributions from religion.

Contributions from Science

Science can give us a better understanding of the environmental impacts of our current agricultural, industrial, and personal practices. In more

*Ian G. Barbour is a professor of religion and the Bean Professor of Science, Technology, and Society emeritus at Carleton College in Northfield, Minnesota.

general terms, the sciences (especially ecology and evolutionary biology) offer a new view of the place of humanity in nature and of our interdependence with other forms of life; this new perception of ourselves will affect our practices if we take it seriously.

Understanding Environmental Impacts

During the 1960s and 1970s, scientists achieved greater knowledge of the impact of our actions on the environment, particularly in the pollution of air, water, and land. They also proposed ways of reducing that impact. Many pollutants were immediately evident and relatively localized, and we have made some progress in regulating and controlling them. In the 1980s and 1990s, scientists have made us aware of more long-term global impacts, such as climate changes and the loss of endangered species. Topsoil around the world is being eroded by intensive agriculture, overgrazing, and deforestation. The world population is growing by eighty million each year. Industrial growth and the consumption of global resources by affluent nations have raised additional questions of long-term sustainability that we have hardly begun to address.[1]

In relation to such scientific evidence, the main task of members of the religious community is to listen to scientists, to become better informed, and to help in spreading the word. But religious participants also have a responsibility to raise questions. They can ask scientists to indicate the assumptions and the uncertainties in their interpretation of data. In the case of global warming, for example, there is virtual unanimity among experts that increases in carbon dioxide and other greenhouse gases will lead to significant warming, though there is a range of estimates of probable temperature changes and their effects on climate, agriculture, and coastal areas. In other cases, there may be greater disagreement among experts. Assessments of risks often depend on assumptions that should be made explicit since they may be influenced by a scientist's ethical or political convictions.

A scientist's recommendation of a particular policy of action is based in part on scientific estimates of the consequences of that policy. But policy recommendations inescapably involve value judgments in weighing the relative importance of diverse kinds of consequence. Furthermore, the main benefits of a new technology or economic policy may accrue mainly to one group of citizens while another group carries the burden of risks or indirect costs, so that issues of distributional justice are at stake. In addition, policy recommendations are the result of comparison among the alternatives that a person thinks are feasible and realistic—which introduces political and economic considerations. Policy decisions thus require ethical analysis and input from the social sciences as well as input from the natural sciences, even when the latter are crucial, as they are in decisions affecting the environment and its sustainability. In short, we need to rely heavily on the work of scientists without expecting them to have the last word on policy choices.[2]

Awareness of Ecological Interdependence

A more indirect contribution of the scientific community to sustainability is an awareness of the interdependence of all forms of life. The study of ecosystems shows the complex interconnections in the web of life. While some ecosystems are quite resilient, others are fragile and vulnerable to the repercussions of human actions. Diversity in the biosphere allows for both stability and adaptation to new conditions. Beyond the study of specific interactions among life forms, ecology has led to a new understanding of our dependence on our environments. Evolutionary biology has also given us a new perception of our place in nature. It has shown our kinship with other creatures. We are united in a common cosmic story that goes back to the early stars, in which were formed the atoms in our brains and in all plants and animals. Life on earth is the story of our family tree of descent from common ancestors. More than 99 percent of the active genes in humans and chimpanzees are identical—though, of course, the remaining 1 percent makes an enormous difference. Scientists see humanity as part of the natural order, even as they acknowledge human capacities not present in other forms of life.[3]

Scientific work in evolutionary biology and neurobiology has been an important influence on recent theologians in their writing about human nature. In the Bible itself, persons are viewed holistically as integrated centers of thinking, feeling, willing, and acting. The Bible portrays the social and bodily character of selfhood. While humanity is given a special status, kinship with other forms of life is also asserted. But under the influence of late Greek thought, the early church came to view a human being as a separate soul temporarily inhabiting a body. This dualism continued in Descartes's distinction between mind and matter as radically different substances. Only humanity was said to have an immortal soul or a rational mind, establishing an absolute line between humans and all other beings. The recent writings of theologians describing a person as an integral psychosomatic being, a bodily self in community, reflects an indebtedness to science as well as a return to biblical roots. As science helps us to see ourselves in a new way, both our attitudes and our actions will be affected.[4]

Joint Contributions from Science and Religion

Let us consider some cases in which similar ideas are found in the scientific and religious communities, even though they are derived from differing sources. Three such themes are (1) a long-term global view, (2) respect for all forms of life, and (3) concern about population growth.

A Long-Term Global View

Degraded land, eroded soil, and decimated fisheries and forests will take many decades to recover. We have been living off biological capital, not biological income. Many of the impacts of our technologies will be felt by future generations. Radioactive wastes from today's nuclear power plants will endanger anyone exposed to them ten thousand years from now. The world of politics, however, takes a very short-term view. Political leaders find it difficult to look beyond the next election. Elected officials receive campaign contributions and heavy lobbying from business and industry, whose attention is directed to short-term profits. Economic calculations give little weight to long-term consequences because a time discount is applied to future costs and benefits.

Scientists, by contrast, typically think in terms of long time spans. Biologists study the history of evolution, and ecologists study populations over many generations. They are keenly aware that population growth in any species puts increasing pressures on the environment and jeopardizes the welfare of future generations. Without the work of scientists, we would never have had the 1987 Brundtland Report of the United Nations, which elaborated the idea of sustainable development and the importance of the needs of future as well as current generations. Scientists from around the world helped to develop the agenda for the 1992 UN Conference on Environment and Development in Rio around the theme of sustainable development.[5]

Religious traditions also take a long-term view. Stewardship requires consideration of the future because God's purposes include the future. The Bible speaks of a covenant from generation to generation "to you and your descendants forever." The land, in particular, is to be held as a trust for future generations. This long time perspective derives from a sense of history and ongoing family and social life, as well as accountability to a God who spans the generations. So it is not surprising that sustainability has been a major theme in statements of the World Council of Churches, several Protestant denominations, and the U.S. Conference of Catholic Bishops.[6]

In addition, both the scientific and the religious communities at their best have advocated a global perspective. Science itself is international. Scientific meetings and journals ignore national boundaries. Scientists have a global viewpoint on environmental and resource problems. They recognize that impacts on the environment at one point have far-reaching repercussions at distant points. CFCs released in repairing an air conditioner in Boston will contribute to the depletion of ozone and the risk of skin cancer in Berlin and Bangkok. Scientists have insisted that an adequate response to global warming can be achieved only by international agreement on the reduction of greenhouse emissions. They have been aware of the destructiveness of nuclear war and have been prominent in the arms control and peace movements and in supporting the United Nations. However, we must acknowledge that scientists are subject to the same nationalistic viewpoints as other citizens, and a large fraction of applied scientists have been working directly or indirectly on military projects.

Our religious traditions have also held up a vision of world community, though they have too often succumbed to intolerance and religious imperialism. The biblical writers affirm our common humanity and assert that we have been made "one people to dwell on the face of the earth." The religious communities of the United States have also been active in working for peace and for support of the United Nations. They have organized famine relief and lobbied for foreign aid and agricultural and technical assistance. Most religious groups have been urging a reduction in military expenditures and have criticized reliance on nuclear weapons. Religious leaders can join scientists in asserting that future threats to the nation's security are likely to be economic and environmental rather than military. Our response to these threats must be global and international rather than narrowly national, for we live in an interdependent world. The end of the Cold War and the emergence of global environmental crises present an unprecedented opportunity for international cooperation.[7]

Respect for All Forms of Life

I noted previously the scientific evidence for the ecological interdependence of all forms of life and the evolutionary understanding of our kinship with other creatures. Human self-interest requires that we consider the impact of our actions on other life forms. But many scientists are critical of a purely anthropocentric viewpoint. They are concerned for the wider web of life and they value other creatures in themselves, not simply for their usefulness to us. Some scientists, such as Aldo Leopold, Loren Eiseley, and Rachel Carson, have given eloquent expression to their experience of unity with and participation in the world of nature. Others respond to the cosmos with a sense of awe and wonder and regard the earth as in some sense sacred. Such personal and reflective responses go beyond science itself, but they provide strong motivation for action to preserve the environment.

Biblical writings also affirm the value of the natural world. The creation story asserts the goodness and unity of the created order. It acknowledges our common status as creatures, even though humanity alone has capacities for personal relationships and moral responsibility that are uniquely in God's image. According to Genesis, humanity is called to till and keep the garden and to exercise careful stewardship of the earth. The covenant after the flood includes all living creatures, and the Deuteronomic law specifies humane treatment of animals and a day of rest for land and animals as well as humans. Many of the Psalms celebrate the beauty and order of nature and its intrinsic value quite apart from its usefulness to us. In subsequent Western history, however, many forces resulted in the increasing alienation of humanity from nature. Medieval thought gave more emphasis to redemption than to creation and referred more often to salvation in a future life than to stewardship of the created order. With the rise of capitalism, nature was viewed as simply a resource for human use. With the growth of technology, human power over nature increased dramatically, and it was assumed that there were no limits to our ability to manipulate it for our own purposes.

But this separation of humanity from nature is now being questioned by theologians as well as scientists. Some of them speak of evolution as God's way of creating, and they see humanity as part of nature, though a unique part. Others seek more frequent celebration of nature in worship and liturgy. Most of the U.S. religious denominations have made statements elaborating the theological and ethical grounds for care of the earth. But local churches and synagogues are only beginning to take these ideas seriously, and much more needs to be done to implement them in practice.[8]

Concern about Population Growth

Population growth is outpacing growth in agricultural production, and it is putting increasing pressure on erodible soils, grazing lands, water, and fuelwood supplies. The average number of children per couple has been slowly falling in most of the third world, but the total population is growing faster than ever because of the large number of young people entering childbearing years. No problem is more urgent, but none is more complex because human reproduction is related to so many variables, including cultural and religious beliefs, family patterns, health services, and economic and educational policies.[9]

Scientists were in the forefront in alerting us to the seriousness of population growth. Ecologists had studied the effects of animal populations on environments of limited carrying capacity, and their work was extended into human ecology. Experts in agronomy and forestry in the third world observed the ways population pressures led to harmful practices. Social scientists observed the consequences of population growth in urbanization and the perpetuation of poverty. They also studied the relation of fertility rates to education, the role of women, and the provision of health services, as well as specific programs for family planning.

At the UN conference in Cairo in 1994, women were for the first time strongly represented in both governmental and nongovernmental delegations. The Program of Action adopted by the conference gave major emphasis to the empowerment of women through access to education, health care, and political and economic equality. The document recognized that birth rates fall when women have more control over their lives. It also called for access to "safe, effective, affordable and acceptable methods of family planning of their choice." Moreover, it pointed to the importance of economic development both as a goal in itself and as a way of encouraging population stability. But the funds pledged to further these goals were quite incommensurate with the magnitude and urgency of the program recommended.[10]

Many church groups were active participants among the nongovernmental organizations at the meetings preparing for Cairo, though the Vatican alone had official standing because of its status as a state. At Cairo, the Vatican and some conservative Muslim nations stood virtually alone in totally opposing contraception as well as abortion. On this issue, the Vatican does not have the support of most Roman Catholics; nine out of ten Catholics in both the United States and Mexico reject the church's teaching, and they practice contraception in almost exactly the same proportion as non-Catholics. Conservative Protestants do not object to contraception, but they have been afraid that the UN would promote abortion as a means of controlling family size, and on those grounds they persuaded the Reagan and Bush administrations to withdraw U.S. support for UN population programs. U.S. funding was restored under the Clinton administration but drastically cut in the 1996 budget agreements, in part out of deference to the religious right.

Mainline Protestant groups have tried to separate the questions of contraception and abortion. They have also insisted that family planning must be included in the wider context of socioeconomic development in the third world and disproportionate consumption and pollution by industrial nations. Most Protestant leaders hold that human sexuality serves not only the goal of procreation but also that of expressing love and unity in marriage, so contraception is an acceptable means of responsible family planning.[11] It is little recognized, however, that in addition to the welfare of individual families, global justice and the integrity of creation are at stake in personal and social decisions about family planning. This is an issue that urgently needs the best joint efforts of the scientific and religious communities.

Contributions from Religion

Let us turn finally to some distinctive contributions of the religious community. The first is commitment to social justice. The second is an alternative vision of the good life that is less consumptive than current patterns in industrial nations.

Commitment to Social Justice

Consumption by industrial nations is responsible for a highly disproportionate share of global resource use, greenhouse gases, and other pollutants. Our dogs and cats are better fed than most of the children in Africa. Many developing nations use their best land to produce export crops in order to pay the interest on their national debts. Poor farmers are forced onto marginal or hillside land that is vulnerable to erosion. Brazil cuts tropical forests mainly for timber export and for agriculture but also to grow feed for cattle for beef sent to American fast-food chains. But the soil is low in nutrients and will support agriculture for only four or five years, so half the Brazilian land cleared for agriculture has already been abandoned. In many third world countries, the pattern of exporting crops is a legacy of colonial imperialism, in which the colonies supplied the raw materials for the industrialization of the colonial powers and provided markets for their manufactured goods. The current international economic order tends to perpetuate the third world's role as a supplier of raw materials and cheap labor.

Within our own nation, the effects of environmental damage fall very unevenly on different groups. The urban poor are exposed to higher levels of air pollution, water pollution, noise, and lead poisoning than citizens with higher incomes, and they have little economic or political power to defend themselves from such risks. Environmental injustice is a product not only of economic differences but also of social and racial inequalities. For example, African-American children from middle-income families are three times as likely to have lead poisoning as children from white families of identical income. Companies looking for a site for a new waste facility are likely to choose an already polluted area, in which they expect to encounter little opposition. There is evidence of discrimination in government decisions as to which toxic waste sites will be cleaned up first, which has led to legislation seeking to enforce equal environmental protection. Another example of unequal risks is the high exposure to pesticides among migrant farm workers, who are predominantly Hispanic.[12]

Many scientists do have a deep personal commitment to social justice. Some scientists, for example, have been active in working for the health and safety of workers and citizens and for greater equity in resource consumption. But I suggest that their commitment to justice did not arise directly from their scientific work, nor can it be supported by scientific evidence. To be sure, in the practice of science itself certain ethical principles are promoted, such as honesty, cooperation, universality, freedom of thought, open communication, and acknowledgment of indebtedness to the work of others. But the actual motives of scientists, like those of all human beings, are mixed. Secrecy, rivalry in seeking recognition, and support of nationalistic or militaristic goals are not uncommon. Moreover, ideals applicable within science do not provide an adequate basis for social ethics in general. I would argue that a scientist's dedication to social justice is a reflection of his or her philosophy of life, whether secular or religious. It is also the product of cultural ideals and of a person's life experiences, exposure to role models, and personal decisions about priorities.[13]

Members of religious communities are equally subject to mixed motives, cultural influences, and life experiences. But in regard to social justice, they do have the support of the Western religious traditions. Ever since the prophets of ancient Israel, the call for justice has been based on the idea of the fundamental equality of all persons in the sight of God. The God of the Bible is portrayed as being on the side of the poor and oppressed. Religious leaders had a major role in the abolition of slavery, the establishment of hospitals, women's suffrage, and more recently, the civil rights movement and the protest against the Vietnam War. Today, environmentalists often neglect social justice, and social activists often neglect the environment, but the religious community can bring these values together.[14] For example, the National Council of Churches has created a Task Force on Eco-Justice. The Catholic bishops' environmental statement cited earlier advocates both solidarity with the poor and respect for nature, and it explores the links between them.

Moreover, the biblical tradition is realistic about human nature and the presence of selfishness and greed in both individuals and institutions. Technology is a source of economic and political power, and power tends to corrupt those who hold it. Abuse of people and abuse of the earth are in large part products of the self-interest of nations, corporations, government bureaucracies, and other organizations. For this reason, religious ethics looks at economic and political institutions as well as individual behavior. Large-scale technologies often result in a concentration of economic and political power, which is difficult to control through democratic processes. Complex technologies are also vulnerable to human fallibility. When humans err with such technologies, or when organizations cut corners on safety measures, the consequences can be tragic, as we saw in the Chernobyl accident, the Bhopal disaster, the Challenger explosion, and the Exxon oil spill. The biblical view of human nature would suggest the need for social regulations to prevent the abuse of power by individuals and institutions in a technological age.

A Less Consumptive Vision of the Good Life

Stronger conservation policies in industrial nations would be a major step toward a more just and sustainable world. Greater efficiency, waste recovery, and cleaner technologies can cut down on both pollution and resource use. But I believe we must go beyond efficiency and look at our patterns of consumption. In our society there are powerful pressures toward the escalation of consumption. By the age of twenty, the average American has already seen 350,000 television commercials. The mass media hold before us the images of a high-consumption lifestyle. Self-worth and happiness are identified with consumer products. Our culture encourages us to try to fill all our psychological needs through consumption. Consumerism is addictive, and like all addictions it involves the denial of its consequences.[15]

I do not think science itself offers us goals for our lives, though it offers the means to achieve some of our goals. But the biblical tradition does offer a vision of the good life, and it is one that is less resource consumptive than prevailing practices. It holds that, after basic needs are met, true fulfillment is found in spiritual growth, personal relationships, and community life. This path is life affirming, not life denying. Religious faith speaks to the crisis of meaning that underlies compulsive consumerism. We should seek a level of sufficiency that is neither ever growing consumption nor joyless asceticism. A vision of positive possibilities and an alternative image of the good life are likely to be more effective than moral exhortation in helping people to turn in new directions. For most people in our nation, restraint in consumption is indeed compatible with personal fulfillment. We can try to recover the Puritan virtues of frugality and simplicity, both in individual lifestyles and in national policies. For the third world, of course, and for low-income families in industrial nations, levels of consumption must rise substantially if basic needs are to be met.[16]

Religious faith can also engender humility and caution in relying on technical fixes for far-reaching social problems. Religion has no quarrel with technology when it serves human needs. Agricultural and medical technologies, for example, are a valid response to the biblical injunction to feed the hungry and care for the neighbor. However, our society has often had an unlimited confidence that all problems can be solved by technological ingenuity, and that there are no limits to our power over nature. But in the last two decades, we have seen that a technical solution to one problem often creates unexpected new problems—like the ozone-depleting effects of CFCs, which no one had anticipated. And religious faith insists that technological improvements are not a substitute for a more humane social order and a more just world.[17]

Conclusion

I have asked why members of the scientific and religious communities should cooperate on issues of population, consumption, and the environment. In the most general terms, they should cooperate because they are human beings concerned about the future of planet Earth. Each of us is a whole person, and we cannot divide our lives into watertight compartments. But we also bring distinctive contributions from the communities to which we belong. Scientists bring specific information about environmental impacts and a new understanding of the relation of humanity to nonhuman nature. Members of both communities share a long-term global view, respect for all forms of life, and concern about population stabilization. Religious communities bring a history of commitment to social justice and a vision of the good life that is less resource consumptive than current patterns in industrial societies. It is to be hoped that we can learn from each other and move together toward the action that is so urgently needed.

Notes

1. Lester Brown et al., *State of the World 1998* (New York: W.W. Norton, 1998); World Resources Institute, *World Resources, 1998–99* (Oxford: Oxford University Press, 1998); Wolfgang Sachs, ed., *Global Ecology: A New Arena of Political Conflict* (London: Zed Books, 1993).

2. Ian G. Barbour, *Technology, Environment, and Human Values* (New York: Praeger, 1980), chaps. 8 and 9, and *Ethics in an Age of Technology* (San Francisco: HarperSanFrancisco, 1993), chaps. 2 and 3.

3. J. Sayer and Timothy Whitmore, eds., *Tropical Deforestation and Species Extinction* (London: Chapman and Hall, 1993); M. L. Reaka-Kudla, D. E. Wilson, and E. O. Wilson, eds., *Biodiversity II: Understanding and Protecting our Biological Resources* (Washington, D.C.: Joseph Henry Press, 1997); Brian Swimme and Thomas Berry, *The Universe Story* (San Francisco: HarperSanFrancisco, 1992).

4. On the relation of scientific to religious views of human nature, see Ian G. Barbour, *Religion and Science: Historical and Contemporary Issues* (San Francisco: HarperSanFrancisco, 1997), chap. 10; Philip Hefner, *The Human Factor: Evolution, Culture and Religion* (Minneapolis: Fortress Press, 1993); Arthur Peacocke, *Theology for a Scientific Age*, enlarged ed. (Minneapolis: Fortress Press, 1993), chap. 12; Malcolm Jeeves, *Human Nature at the Millennium* (Grand Rapids, Mich.: Baker Books, 1997).

5. World Commission on Environment and Development, *Our Common Future* (Oxford: Oxford University Press, 1987); *The Global Partnership for Environment and Development: A Guide to Agenda 21* (New York: United Nations Publications, 1992).

6. United States Catholic Conference, "Renewing the Earth" (1991); Presbyterian Church USA, "Keeping and Healing the Creation" (1989); Evangelical Lutheran Church in America, "Caring for Creation" (1991). On the World Council of Churches, see Welsey Granberg-Michaelson, *Redeeming the Creation: The Rio Earth Summit—Challenges for the Churches* (Geneva: WCC, 1992).

7. Norman Myers, *Ultimate Security: The Environmental Basis of Political Stability* (New York: W.W. Norton, 1993).

8. James A. Nash, *Loving Nature: Ecological Integrity and Christian Responsibility* (Nashville: Abingdon Press, 1991); Max Oelschlaeger, *Caring for Creation: An Ecumenical Approach to the Environmental Crisis* (New Haven, Conn.: Yale University Press, 1994); Larry Rasmussen, *Earth Community, Earth Ethics* (Maryknoll, N.Y.: Orbis, 1996).

9. Laurie Mazur, ed., *Beyond the Numbers: A Reader on Population, Consumption, and the Environment* (Washington, D.C.: Island Press, 1994); Benedicta Musembi and David E. Anderson, *Religious Communities and Population Concerns* (Washington, D.C.: Population Reference Bureau, 1994).

10. United Nations, *Report of the International Conference on Population and Development: Cairo, 1994* (New York: United Nations, 1995); "Link-

ing Population and Development in Cairo," *Ecojustice Quarterly* 15, no. 1 (Winter 1995).

11. Susan Power Bratton, *Six Million and More: Human Population Regulation and Christian Ethics* (Louisville: Westminster/John Knox, 1992).

12. Robert Bullard, ed., *Confronting Environmental Racism* (Boston: South End Press, 1993); Laura Westra and Peter Wenz, eds., *Faces of Environmental Racism: Confronting Issues of Global Justice* (Lanham, Md: Rowman and Littlefield, 1995).

13. Barbour, *Ethics in an Age of Technology*, chap. 2.

14. Donald B. Conroy and Rodney L. Petersen, eds., *Earth at Risk: Advancing Dialogue Between Religion and Science* (Amherst, N.Y.: Humanity Books, 1999); Peter Bakken, Joan Gibb Engel, and J. Ronald Engel, eds., *Justice, and Christian Faith: A Critical Guide to the Literature* (Westport, Conn.: Greenwood Press, 1995).

15. Alan Durning, *How Much Is Enough? The Consumer Society and the Future of the Earth* (New York: W.W. Norton, 1992); Paul Wachtel, *The Poverty of Affluence: A Psychological Portrait of the American Way of Life* (Philadelphia: New Society Publishers, 1989); Lester Milbraith, *Envisioning a Sustainable Society* (Albany: State University of New York, 1989).

16. James A. Nash, "Toward the Revival and Reform of the Subversive Virtue: Frugality," *Annual of the Society of Christian Ethics* (1995), 137–60; James Martin-Schramm, "Population-Consumption Issues: The Debate in Christian Ethics," in ed. Dieter T. Hessel, *Theology for Earth Community: A Field Guide,* (Maryknoll, N.Y.: Orbis, 1996); Wesley Granberg-Michaelson, "An Ethics of Sustainability," *Ecumenical Review* 42 (Jan. 1991): 120–30.

17. Robert Stivers, *Hunger, Technology, and Limits to Growth* (Minneapolis: Augsburg Press, 1984); Herman Daly and John Cobb, Jr., *For the Common Good: Redirecting the Economy toward Community, the Environment, and a Sustainable Future* (Boston: Beacon Press, 1989).

Part II

SCIENCE

Overview of Scientific Perspectives

Audrey R. Chapman

What precisely has been the impact of human population growth, technology, and consumption patterns on the environment, both specific ecological systems and the planet as a whole, and what kinds of future trends are we likely to have? To what extent is the growth in human population, affluent societal consumption patterns in the industrial countries, conspicuous consumption among the upper class in less economically developed countries, and associated economic activity placing great stress on the earth's environmental support systems? What is the significance of the various indications that nature is in trouble, such as land degradation in some areas, severe pollution damage, loss of biological diversity, falling water levels, declining bird and wildlife populations, and record rises in global temperatures? Are they temporary and reversible or signs of something more systemic and problematic? Have we already reached or exceeded the limits of the earth's carrying capacity? What is the number of individuals that can be supported without degrading the natural, cultural, and social environments? How do we begin to project future trends and address these issues? This section explores these complex and significant issues. It also considers the extent to which science can provide precise answers to these questions.

Trends

Population figures attest to the ever escalating increase in human numbers. Demographers estimate there were approximately 5 million people ten thousand years ago. This small population increased to some 250 million by the birth of Christ. Subsequent milestones were 500 million in the fourteenth century, 1 billion in 1830, 2 billion in 1930, and 5 billion in 1988.[1] Human population is growing by geometric proportions. While it took some fourteen hundred years for the population of 250 million at the beginning of the common era to double, the first billion doubled into 2 billion people in the space of one century, and within less than a half century 2 billion became a population of 4 billion. Despite declines in rates of growth during the past generation, the actual number of people being added each year is currently at an all-time high, approaching 90 million people annually.[2]

Projections about the future size of the human population vary, depending on assumptions about fertility and mortality rates. The dilemma is that demographers either do not have sufficient data or do not fully understand the determinants of human population levels. Societal fertility rates depend on an aggregate of individual women's family sizes, and very little research has been done on the parameters of decision making at the family level. In the modern era, mortality rates have been falling in response to the introduction of public health practices, sanitation, and modern medicine, first in the industrialized countries and then, in the twentieth century, in poorer countries. Given current economic stresses, it is difficult to know whether the poorest countries will be able to sustain or improve investments in these critical areas. Recently, fertility rates have declined in many poorer countries, reflecting smaller average family sizes, but researchers have not been able to explain satisfactorily why this is happening.

Specific projections, particularly those extrapolating from current trends, are daunting. If the world population were to continue to grow at 1990 rates, by 2150, some 150 years from now, it will approach 694 billion, an increase of 130-fold.[3] Fortunately, this is unlikely to happen. Updated United Nations estimates, which appear in "World Population Projections to 2150," range from a "low" of 7.3 billion people to a "high" of about 10.7 billion people in the middle of the twenty-first century. The new population figures suggest that the world's population, calculated to be approximately 5.9 billion in 1998, will reach 8.9 billion by 2050 and then stabilize (by means of a balancing of birth and death rates) just after the year 2200 at 10.73 billion. The lower population projections reflect data showing a decline in fertility from the anticipated 3.1 children per woman to 2.7 children per woman, higher than anticipated death rates in some areas, and declining life expectancy in Eastern Europe and the former Soviet Union.[4]

Nonetheless, population growth remains a serious problem. Even the most optimistic scenarios anticipate something approaching a doubling of the current human population. Because women of reproductive age make up a large and growing percentage of the population in many developing countries, there will be a "forward thrust" in population expansion even if the average woman continues to have fewer children. Moreover, there is no assurance that the lower population growth rate will be maintained. Some experts project that, in the absence of lower levels of consumption in affluent countries and new economic models in the adoption of developing countries, a five- to tenfold increase in global economic activity is likely to be necessary to meet the needs and aspirations of a population of 10 billion people.[5] While some economists believe that this scale of economic expansion may be theoretically feasible, it clearly will not be environmentally sustainable.

Even the lower projections of population increase over the next century envision increases in population that would place considerably greater, potentially overwhelming, stress to provide an adequate supply of food, which has profound implications for maintaining the integrity of ecosystems. Multiplication of human numbers is likely to cause great human misery by worsening already alarming rates of hunger, poverty, and migration. Already between one and two billion humans are chronically malnourished, the largest number of hungry humans ever recorded. These numbers are likely to accelerate as the global population increases. While improved technology can assist in more effective management and use of resources, it cannot ensure adequate food supplies for future generations unless population growth is curtailed.

The impact of population growth on the environment, on development, and on the quality of life is compounded by the fact that the greatest increase is taking place in poor countries less able to provide services and infrastructure for growing numbers. It is projected that 95 percent of the anticipated population increase will take place in less developed regions. Moreover, as data in the *Report of the International Conference on Population and Development* indicate, the global figures mask large disparities in demographic trends between the more affluent industrialized countries and the poorer countries. Between 1985 and 1990, for example, the average number of children born per woman ranged from 1.3 children per woman in Italy to 8.5 children per woman in Rwanda, and life expectation at birth ranged from an estimated 41 years in Sierra Leone to 78.3 years in Japan. In several countries with annual rates of population growth of 3 percent or more in the 1980s—Kenya, Zambia, Nigeria, Ghana, most of the Arab countries, Pakistan, and a few Latin American countries—population doubling will occur in about twenty years.[6]

These differentials have significant implications for the prospects of sustainable and equitable development. Rapid poverty-related population growth is a major cause of land degradation in many areas and limits the possibilities for achieving sustainable economic development and improving the quality of life. Widespread poverty is often accompanied by unemployment, malnutrition, illiteracy, low status of women, exposure to environmental risks, and limited access to social and health services, including reproductive and family planning services. All of these factors in turn contribute to high levels of fertility, morbidity, and mortality, as well as to pressures on the environment and low economic activity. Conversely, efforts to slow down population growth, reduce poverty, reduce unsustainable consumption and production patterns, and improve environmental protection have a mutually reinforcing effect.

In assessing human pressure on the environment, it is important to note that high rates of natural resource consumption and pollution, primarily in affluent countries, have at least as much impact as demographic pressures. Environmental degradation results both when people have too little and when people have too much. Members of consumer societies are responsible for a disproportionate share of the global environmental challenges now facing humanity. According to the essay by William Rees, most so-called advanced countries are running massive ecological deficits with the rest of the planet. Put another way by Rees, most industrialized countries, with the exception of Canada, are overpopulated in ecological terms—that is, they could not sustain themselves at current material standards if forced by changing circumstances to live on their remaining endowments of domestic natural capital. The following figures attest to the scale of the global imbalance: Each year the richest 20–25 percent of the world's population consumes approximately 75–80 percent of all resources and produces most of the planet's pollution and waste. The wealthy consume on average three times their fair share of sustainable global output. For example, two-thirds of the emissions of the carbon dioxide gases considered to be responsible for climate change come from affluent countries. It is estimated that each child born today in a Northern Hemisphere country is likely to consume and pollute twenty to thirty times as much as a child born in the south. Historically, the industrialized nations have also been responsible for most of the pollutants that have accumulated in the biosphere.[7]

Moreover, Americans rank near the top in most categories of resource use and waste, even when compared to other European countries and Japan. As the essay by Michael Brower and Warren Leon shows, American consumption habits impose a severe cost on the environment. The average American uses twice as much fossil fuel as the average resident of Great Britain and two and a half times as much as citizens of Japan. U.S. per capita emissions of carbon dioxide are twice as high as those of Germany and three times as high as those of Switzerland and Italy; U.S. energy and water use per capita are about twice that of Germany and three or four times that of Switzerland and Italy; each person in the United States generates thirteen times as much waste as each person in Germany and twenty-one times as much as each person in Switzerland.[8] Reflecting our long-standing love affair with the automobile, Americans own more cars and drive them longer distances than citizens of other nations while neglecting other more energy-efficient modes of transportation. Our agricultural system is extremely energy intensive, accounting for nearly half of U.S. energy consumption, and uses large quantities of pesticides and fertilizers. Moreover, Americans tend to have much less interest in adopting energy-saving and pollution-reducing technologies than citizens of other industrialized countries do.

Unless affluent societies exhibit a willingness to reduce consumption and adopt new economic arrangements and technologies, poorer countries are unlikely to reconsider their development models and goals. If developing countries, particularly those with large populations like India and China, seek to improve living standards through emulating Western economic institutions and technologies, this will greatly accelerate stress on the earth's life-support systems. The current situation is not hopeful. Given the training and orientation of the leadership of most poor countries and the influence of international economic institutions promoting capitalist market economies, the development of poor and middle-income countries seems destined to repeat the environmentally unsustainable trajectory of the industrialized nations. Moreover, the gap between the affluent minority and the very poor majority characteristic of the global division of resources is replicated within country after country: a privileged minority engages in ostentatious consumption while the majority of people are mired in dire poverty.

Yet the people of affluent countries, Americans in particular, have been very reluctant to deal with the impact of their consumption patterns on the environment. The concluding documents of both the 1992 Rio Conference (the United Nations Conference on Environment and Development) and the 1994 Cairo (International) Conference on Population and Development were notable in their virtual silence on this issue. Far from being accidental, the unbalanced treatment of population growth and overconsumption as environmental problems in those documents resulted from strenuous lobbying by developed countries.

While northern countries may be able to impose their perspectives at international conferences, the cost is great. It undermines the credibility of the agreements reached, and, perhaps more to the point, it delays making the hard choices that are inevitably ahead.

Some analysts argue that the entrenched inequality between rich and poor countries, which recent surveys show is increasing,[9] underlies the population-consumption-environmental crisis. A. J. McMichael contends, for example, that rapid, poverty-related, population growth and land degradation in poor countries and excessive consumption of energy and materials, with high production of wastes, in rich countries are both manifestations of this inequality. McMichael foresees three possible adverse outcomes of these patterns: (1) exhaustion of various nonrenewable materials, (2) toxic contamination of localized environments, and (3) impairment of the stability and productivity of the biosphere's natural system. Of the three, the most attention, at least in the past, was paid to the first. Currently, we are beginning to document the effects of localized exhaustion of resources and chemical pollution. The third would constitute by far the most serious potential threat, and McMichael and others claim that some of today's global environmental changes may portend that third outcome.[10]

Contributions and Limitations of Science

As the 1990 "Open Letter to the Religious Community" states, efforts to safeguard the environment require a wider and deeper understanding of science and technology: "If we do not understand the problem, it is unlikely we will be able to fix it."[11] Certainly, science provides an essential starting point for assessing the state of the environment and our options for solving major problems.

Ian Barbour's essay in part I outlines some of the potential contributions of science. First, in the past thirty years scientists have achieved greater knowledge of the impact of our actions on the environment and ways to reduce and ameliorate those impacts. Second, the scientific study of ecosystems has promoted an awareness of the interdependence of all life forms and our kinship with other creatures. Ecology has documented the importance of diversity within the biosphere and ecosystems for stability and adaptation to new conditions. It has also underscored the vulnerability of many life forms and the scope of extinctions from human activities. Scientists therefore realize that population growth in any species puts increasing pressure on the environment and jeopardizes the welfare of future generations, both of humanity and of other creatures. Third, according to Barbour, science, like religion, offers a long-term and global view. Many scientists therefore are aware of and concerned about the impact of our technologies on future generations. Degraded land, polluted waterways, and decimated fisheries will take many decades to recover. Human-induced climate change will affect habitats and the possibilities for preserving life indefinitely. Current trends in population growth and continued living off biological capital will present future generations with increasingly bleak options.

However, there is little likelihood that science will provide conclusive answers, at least in the near term, to critical issues related to the subject matter of this volume. Despite the best efforts of demographers, long-term population projection is still more of an art than a science. Scientists can provide indicators of sustainability, as William Rees does in his essay, but it is more complex to undertake actual calculations of ecological carrying capacity to determine how large an area of productive land is needed to sustain a defined population indefinitely. Many, and probably most, scientists concur that there are clear limits to the ability of a finite planet to sustain indefinitely the anticipated load of an expanding human population and economy without progressively impairing the integrity of relevant ecosystems. Nevertheless, science cannot currently determine conclusively whether we have already reached or will soon reach the limits of the earth's carrying capacity.

Part of the problem is the difficulty of measuring the interrelationships between human populations and the environment. The essays by Joel Cohen and William Rees show that ecological notions of carrying capacity that are appropriate for nonhuman populations cannot easily be applied to human populations. Once the critical element of human choice is factored in, as indeed it must be, the complexity of the task increases significantly. As Cohen underscores, the answer to the question of how many people the earth can support depends as much on social, cultural, economic, and political choices as it does on constraints imposed by nature. The average level of material well-being, the distribution of resources, and the nature of technology matter greatly in determining the per capita impact on the planet. And these factors are likely to change. Research on many of the most significant environmental issues therefore depends on making assumptions and extrapolating data from dynamic trends.

Moreover, in many instances there are no agreed upon methodologies for dealing with these dynamic and complex issues. Estimates of how many people the earth can support differ significantly because of differences in methodology, as well as divergent assumptions about likely social, cultural, economic, and political choices that will affect natural resource consumption. Cohen indicates that current estimates of the earth's maximum supportable human population use one of six methods, none of which is fully adequate and several of which are very problematic. Rees offers another approach based on calculating an area-based sustainability indicator that he terms an "ecological footprint," but he acknowledges that these calculations are more complicated than they appear from the basic concept. Moreover, his conclusion—that the equivalent of five additional earths would be needed for a population between ten and eleven billion just to maintain the present rate of ecological decline—will be unacceptable to many.

The situation is made even more complex by disagreements among scientists and policy makers regarding the standard of proof. There are many landmarks to suggest that we have overshot the optimal scale of human economy but little hard data to conclusively prove it. Scientists may interpret the same data differently, depending on whether they tend toward caution or optimism about the future, and may also disagree in their recommendations. Many of those reluctant to change argue that there is insufficient evidence to warrant major societal investments to protect the environment. Others respond that irrefutable evidence of human-caused global warming or species extinction may not be available until it is too late to prevent those disasters. Religious communities, many of which have a strong sense of responsibility to the Creator and to future generations, can play an important role in this debate by supporting the need to err on the side of caution.

Notes

1. A. J. McMichael, *Planetary Overload: Global Environmental Change and the Health of the Human Species* (Cambridge, England: Cambridge University Press, 1993), p. 112.

2. *Report on the International Conference on Population and Development, Cairo, 5–13 September 1994* (New York: United Nations, 1995), p. 32.

3. Joel E. Cohen, "How Many People Can the Earth Support?" *The Sciences* (November–December 1995): 19.

4. Barbara Crossette, "World Is Less Crowded Than Expected, the U.N. Reports," *New York Times*, November 19, 1996, p. 2.

5. Gretchen C. Daily and Paul R. Ehrlich, "Population, Sustainability, and Earth's Carrying Capacity," *BioScience* 42, no. 10 (November 1992): 761–71, cited in James B. Martin-Schramm, "Population-Consumption Issues: The State of the Debate in Christian Ethics," in Dieter T. Hessel, ed., *Theology for Earth Community: A Field Guide* (Maryknoll, N.Y.: Orbis, 1996), p. 133.

6. *Report of the International Conference on Population and Development,* pp. 17, 20, pars. 3.13, 3.14, 3.23.

7. World Wildlife Fund, *Population and the Environment* (Gland, Switzerland: WWF, 1994), p. 7.

8. Alan Thein Durning, "Consumption: The Neglected Variable in the Population and Environment Equation," *Sojourners* (August 1994), p. 22.

9. Barbara Crossette, "U.N. Survey Finds World Rich-Poor Gap Widening," *New York Times,* July 15, 1996, p. 1.

10. McMichael, *Planetary Overload,* p. 7.

11. "An Open Letter to the Religious Community," January 1990. This letter is available from the Science Office of the National Religious Partnership for the Environment, P.O. Box 9105, Cambridge, MA 02238.

World Population Projections to 2150

The Population Division of the Department of Economic and Social Affairs at the United Nations Secretariat prepares biennially the official United Nations population estimates and projections for countries, urban and rural areas, and major cities for all countries and areas of the world. The latest revision, World Population Prospects: The 1996 Revision, *refers to the period 1950–2050. Realization of the full consequences of changes in fertility and mortality on population growth requires a longer time frame. Therefore, periodically the Population Division prepares population projections for an extended projection horizon, in this case to 2150.*

This essay (available at http://www.undp.org/popin/wdtrends/execsum.htm) is an unedited excerpt of the report of a study by the Population Division, Department of Economic and Social Affairs, at the United Nations Secretariat, completed in February 1998. The study was supported in part by a grant from The David and Lucile Packard Foundation.

Executive Summary

The long-range population projections presented here, prepared by the United Nations Population Division, cover the period from 1950 to 2150. A total of seven projections for each of the eight major areas of the world are considered in this report. The variants are distinguished by their assumptions regarding future scenarios in total fertility rates. The range of potential demographic outcomes underscores the difficulty in focusing on any particular scenario and also highlights the critical importance of current policies and actions for the long-range future of the world population.

The seven main conclusions from these long-range population projections are:

- According to the medium-fertility scenario, which assumes fertility will stabilize at replacement levels of slightly above two children per woman, the world population will grow from 5.7 billion persons in 1995 to 9.4 billion in 2050, 10.4 billion in 2100, and 10.8 billion by 2150, and will stabilize at slightly under 11 billion persons around 2200.
- Although the high and low fertility scenarios differ by just one child per couple, half a child above and half a child below replacement fertility levels, the size of the world population in 2150 would range from 3.6 billion persons to 27.0 billion.
- If fertility rates were to stay constant at 1990–1995 levels for the next 155 years, the world in 2150 would need to support 296 billion persons.
- If all couples of the world had begun to bear children at the replacement-fertility level in 1995 (about 2 children per couple), the growth momentum of the current age structure would still result in a 67 percent increase in the world population, to 9.5 billion by 2150.
- The future will see a continued geographical shift in the distribution of the world population as the share living in the currently more developed regions will decrease from 19 to 10 percent between 1995 and 2150.
- Declining fertility and mortality rates will lead to dramatic population aging. In the medium-fertility scenario, the share aged 60 years or above will increase from 10 to 31 percent of the world population between 1995 and 2150.
- The ultimate world population size of nearly 11 billion persons, according to the medium fertility scenario of these projections, is 0.7 billion persons fewer than previously published by the United Nations in 1992, mainly due to larger-than-expected declines in fertility in many countries.

Future Size of the World Population

The medium-fertility scenario assumes that total fertility rates will ultimately stabilize by the year 2055 at replacement levels, which are slightly above two children per woman. The medium-fertility scenario, which lies in the center of the projection scenarios, indicates that the world population will reach 9.4 billion by 2050, 10.4 billion by 2100, and 10.8 billion by 2150. The population of the world will ultimately stabilize at just under 11 billion persons around 2200.

The fertility gap separating the high- and the low-fertility scenarios is about one child. The high-fertility scenario assumes that total fertility rates will converge by 2050 to between 2.50 and 2.60 children per woman and the low-fertility scenario assumes that the total fertility rates will eventually stabilize at levels between 1.35 and 1.60 children per woman. According to the high-fertility scenario, world population will grow to 11.2 billion persons by 2050, 17.5 billion by 2100, and 27.0 billion by 2150. The low-fertility scenario is in sharp contrast and shows a world population increasing to 7.7 billion persons by 2050 but then declining to 5.6 billion in 2100 and eventually falling to 3.6 billion by 2150.

There are two intermediate scenarios, the high/medium- and low/medium-fertility scenarios, which assume that fertility rates will follow the high and the low patterns respectively until about 2025, after which they converge to 10 percent above replacement level (in the high/medium-fertility extension) and 10 percent below replacement level (in the low/medium-fertility extension). According to these two scenarios, the world population in 2150 would reach 18.3 billion and 6.4 billion, respectively.

The constant-fertility scenario presents the results of future world population growth if fertility rates were to remain at 1990–1995 levels through the year 2150. The results highlight the unsustainability of the current situation: the world population under the constant-fertility scenario would reach 296 billion by the year 2150.

An additional analytical scenario demonstrates the vital role of age structure in influencing long-term population growth. Even if all couples in the world had begun in 1995 to bear children at the replacement-fertility level (instant replacement fertility scenario, which is roughly two children per couple), the built-in growth momentum of the population age structure will mean that the population of the world would continue to grow to 9.5 billion by the year 2150—a 67 percent increase from 1995.

The ultimate world population size of nearly 11 billion persons according to the medium fertility scenario of these United Nations long-range projections is six percent lower than the 11.6 billion previously projected by the United Nations in 1992. The reduction of about 0.7 billion persons is mainly due to larger declines in the fertility rates in the developing countries than previously projected.

Geographical Distribution of the World Population

The growth of the major areas of the world is far from homogenous. According to the results of the medium-fertility scenario, population growth will continue in all major areas except Europe. The population of Africa will nearly quadruple over the 155 year period—increasing from 0.7 billion persons in 1995 to 2.8 billion in 2150. Growth is also projected for Asia, with China growing from 1.2 billion persons to 1.6 billion, India from 0.9 billion persons to 1.7 billion, and the rest of Asia from 1.3 billion persons to 2.8 billion. The population of Latin America and the Caribbean is projected to rise from 477 million persons in 1995 to 916 million in 2150.

Northern America's population in 1995 is estimated at 297 million persons and it is expected to increase to 414 million by 2150. Oceania is expected to increase from 28 million persons to 51 million. Europe is the only major area whose population is projected to decline over time. In 1995, Europe's population stood at 728 million persons. By 2150, it is projected to fall to 595 million persons—a decline of 18 percent over 155 years.

The different growth paths of the major areas of the world will lead to a substantial redistribution of the world population across the globe. The medium-fertility scenario projects a decline in the proportion of the world population living in Europe and Northern America and an increase in the proportion found in Africa and several other parts of the world that are today categorized as less developed regions. In 1950, the population of Europe was more than twice that of Africa. However, by 2150, the population of Europe is projected to be one-fifth the size of Africa according to the medium-fertility scenario.

Age Structure of World Population

The growth in the size of the world population is matched by a large shift in the age structure of the world population. The medium-fertility scenario indicates that the median age of the world population will rise from 25.4 years in 1995, to 36.5 years in 2050, to 42.9 years by 2150.

The share of the world's population under 15 years of age will decline from 31 percent in 1995 to 17 percent by 2150. In contrast, the percentage of the population of the world aged 60 or above will increase rapidly from 9 percent in 1995 to 30 percent in 2150. Among the elderly age groups, it is the oldest old—those aged 80 or over—that will increase the most rapidly over time. According to the medium-fertility scenario, the number aged 80 or over will grow from 61 million in 1995 to 320 million in 2050 and 1,055 million by 2150.

Conclusion

While it is certain that the world population will continue to grow significantly in the medium term, there is less certainty in the longer term. The medium-fertility scenario projects that world population could reach 10.8 billion persons by 2150 and would ultimately stabilize at nearly 11 billion persons around 2200. However, the low- and high-fertility scenarios put a large band around the numbers—from 3.6 billion persons in 2150 to 27.0 billion in 2150. This range of potential demographic outcomes underscores the difficulty in focusing on any particular scenario and also highlights the critical importance of current policies and actions for the long-range future of the world population.

Population Growth and Earth's Human Carrying Capacity

Joel E. Cohen*

Earth's capacity to support people is determined both by natural constraints and by human choices concerning economics, environment, culture (including values and politics), and demography. Human carrying capacity is therefore dynamic and uncertain. The element of human choice is not captured by ecological notions of carrying capacity that are appropriate for nonhuman populations.

Scientific uncertainty about whether and how the earth will support its projected human population has led to public controversy: Will humankind live amid scarcity or abundance or a mixture of both?[1]

The Past and Some Possible Futures

Over the past two thousand years, the annual rate of increase of global population has grown about fifty-fold from an average of 0.04 percent

*Joel E. Cohen is the Abby Rockefeller Mauzé Professor of Populations at Rockefeller University and Columbia University. This essay contains excerpts from "Population Growth and Earth's Human Carrying Capacity," *Science* 269 (July 21, 1995): 341–46; and from "How Many People Can the Earth Support?" *The Sciences* 35, no. 6 (November–December 1995): 18–23. Copyrights © 1995 American Association for the Advancement of Science, reprinted by permission of the American Association for the Advancement of Science; and © 1995 Joel E. Cohen. The author acknowledges with thanks U.S. National Science Foundation grant BSR92-07293 and the hospitality of Mr. and Mrs. William T. Golden, Jesse H. Ausubel, Danny J. Boggs, Griffith M. Feeney, Richard B. Gallagher, Shiro Horiuchi, and Robert M. May, who reviewed and improved previous drafts.

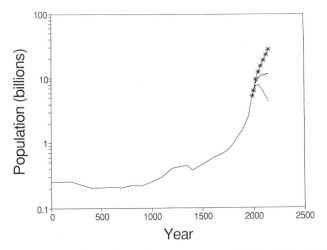

Figure 1. Recent world population history A.D. 1–1990 (solid line)[37] and 1992 population projections of the United Nations[10] from 1990 to 2150. Population growth was faster than exponential from about 1400 to 1970. Asterisks, dashes, and dots indicate high, medium, and low projections, respectively.

per year between A.D. 1 and 1650 to its all-time peak of 2.1 percent per year around 1965–70.[2] The growth rate has since declined haltingly to about 1.4 percent per year (figure 1).[3]

Human influence on the planet has increased faster than the human population. For example, while the human population more than quadrupled from 1860 to 1991, human use of inanimate energy increased from 1 billion megawatt-hours per year to 93 billion megawatt-hours per year (figure 2). In the minds of many, human action is linked to an unprecedented litany of environmental problems, some of which affect human well-being directly.[4] As more humans contract the viruses and other pathogens of previously remote forests and grasslands, dense urban populations and global travel increase opportunities for infections to spread.[5] The wild beasts of this century and the next are microbial, not carnivorous.

Along with human population, the inequality in the distribution of global income has grown in recent decades.[6] In 1992, 15 percent of people in the world's richest countries enjoyed 79 percent of the world's income.[7] Economic contrasts are compounded by cultural ones. In every continent, in giant city-systems, people increasingly come into direct contact with others who vary in culture, language, religion, values, ethnicity, and socially defined race, and who share the same space for social, political, and economic activities.[8] The resulting frictions are evident in all parts of the world.

As of 1999, the world has about 6 billion people. The population would double in forty-nine years if it continued to grow at its present 1.4 percent per year, though that is not likely. The population of less developed regions is growing at 1.7 percent per year, while that of more developed regions is growing at 0.1 percent per year.[9]

The future of the human population, like the future of its economies, environments, and cultures, is highly unpredictable. The United Nations regularly publishes projections with a range from high to low (figure 2). In 1992, its high projection assumed that the worldwide average number of children born to a woman during her lifetime at current birth rates (the total fertility rate, or TFR) would fall to 2.5 children per woman in the twenty-first century; in that scenario, the population would grow to 12.5 billion by 2050.[10] Its 1992 low projection assumed that the worldwide average TFR would fall to 1.7 children per woman; in that case, the population would peak at 7.8 billion in 2050 before beginning to decline.

There is much more uncertainty about the demographic future than such projections suggest.[11] At the high end, the TFR in the less developed countries today, excluding China, is about 3.8 children per woman; that region includes 3.5 billion people. Unless fertility in the less developed countries falls substantially, global fertility could exceed that assumed in the UN's high projection. At the low end, the average woman in Germany now has about 1.3 children, and in Italy and Spain

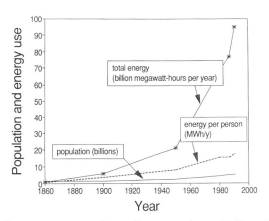

Figure 2. Inanimate energy use from all sources from 1860 to 1991: aggregate (solid line with asterisks)[38] and per person (dashed line). Global population size is indicated by the solid line.

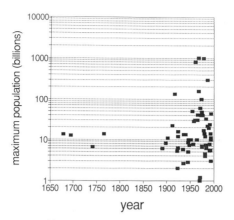

Figure 3. Estimates of how many people earth can support, by the data at which the estimate was made. When an author gave range of estimates or indicated only an upper bound, the highest number is plotted here.[39]

1.2. Fertility could fall well below that assumed in the UN's low projection.

Can the earth support the people projected for 2050? If so, at what levels of living? In 1679, Antony van Leeuwenhoek (1632–1723) estimated that the maximum number of people the earth can support is 13.4 billion.[12] Many more estimates of how many people the earth can support followed (figure 3).[13] The estimates have varied from < 1 billion to >1,000 billion. Estimates published in 1994 alone ranged from < 3 billion to 44 billion.[14] Since 1679, there has been no clear increasing or decreasing trend in the estimated upper bounds. The scatter among the estimates has increased with the passage of time. This growing divergence is the opposite of the progressive convergence that would ideally occur when a constant of nature is measured. Such estimates deserve the same profound skepticism as population projections. They depend sensitively on assumptions about future natural constraints and human choices.

Many authors give both a low estimate and a high estimate. Considering only the highest number given when an author stated a range, and including all single or point estimates, the median of sixty-five upper bounds on human population was 12 billion. If the lowest number given is used when an author stated a range of estimates, and all point estimates are included otherwise, the median was 7.7 billion. This range of low to high medians, 7.7–12 billion, is very close to the range of low and high UN projections for the population in 2050: 7.8–12.5 billion.

A historical survey of estimated limits is no proof that limits lie in this range. It is merely a warning that the human population is entering a zone where limits on the human carrying capacity of earth have been anticipated and may be encountered.

Methods of Estimating Human Carrying Capacity

Estimates of earth's maximum supportable human population are made with one of six methods, apart from those that are categorical assertions without data. First, several geographers divided earth's land into regions, assumed a maximum supportable population density in each region, multiplied each assumed maximal population density by the area of the corresponding region, and summed over all regions to get a maximum supportable population of earth. The assumed maximum regional population densities were treated as static and were not selected by an objective procedure.

Second, some analysts fitted mathematical curves to historical population sizes and extrapolated them into the future.[15] As the causal factors responsible for changes in birth rates and death rates were and are not well understood, there has been little scientific basis for the selection of the fitted curves.

Third, many studies focused on a single assumed constraint on population size, without checking whether some other factors might intervene before the assumed constraint comes into play. The single factor most often selected as a likely constraint is food.[16] In 1925, the German geographer Albrecht Penck stated a simple formula that has been widely used:[17]

$$\text{population that can be fed} = \text{food supply} / \text{individual food requirement}. \qquad [1]$$

This apparently objective formula can lead to extremely different estimates of maximum supportable population because it depends on estimates of the food supply and of individual requirements. The food supply depends on areas to be planted and watered, choice of cultivars, yields, losses to pests and waste, cultural definitions of what constitutes acceptable food, and random fluctuations of weather. Individual requirements depend on the calories and protein consumed directly, as well as on nutrients used as animal fodder.[18] Besides food, other factors proposed as sole constraints on human numbers include energy, biologically accessible nitrogen, phosphorus, fresh water, light, soil, space, diseases, waste disposal, nonfuel minerals, forests, biological diversity, and climatic change.

Fourth, several authors reduced multiple requirements to the amount of some single factor. For example, in 1978 Eyre reduced requirements for food, paper, timber, and other forest products to the area of land required to grow them.[19] Other factors that cannot be reduced to an area of land, such as water or energy, are sometimes recognized indirectly as constraints on the extent or productivity of cultivable land. The authors who combined different constraints into a single resource assumed that their chosen resource intervened as a constraint before any other factor.

Fifth, several authors treated population size as constrained by multiple independent factors. For example, Westing, in 1981, estimated the constraints on population imposed independently by total land area, cultivated land area, forest land area, cereals, and wood.[20] Constraints from multiple independent resources are easily combined formally. For example, if one assumes, in addition to a food constraint, a water constraint

population that can be watered =
     water supply / individual water requirement   [2]

and if both constraints [1] and [2] must be satisfied independently, then

population that can be fed and watered =
   minimum of {food supply / individual food requirement
   or water supply / individual water requirement}.   [3]

This formula is an example of the law of the minimum proposed by the German agricultural chemist Justus Freiherr von Liebig (1803–73).[21] Liebig's law of the minimum asserts that, under steady-state conditions, the population size of a species is constrained by whatever resource is in shortest supply.[22] Liebig's law has serious limitations when it is used to estimate the carrying capacity of any population. If different components of a population have heterogeneous requirements, aggregated estimates of carrying capacity based on a single formula will not be accurate; different portions of the global human population are likely to have heterogeneous requirements. In addition, Liebig's law does not apply when limiting factors fluctuate, because different factors may be constraining at different times; an average over time may be misleading. Liebig's law assumes that the carrying capacity is strictly proportional to the limiting factor (within the range where that factor is limiting); strictly linear responses are not generally observed.[23] Liebig's law assumes no interactions among the inputs; independence among limiting factors is not generally observed. (For example, equation [3] neglects the possibility that changes in the water supply may affect the food supply through irrigation.) Liebig's law assumes that adaptive responses will not alter requirements or resources during the time span of interest; economic history (including the inventions of agriculture and industry) and biological history (including the rise of mutant infections and the evolution of resistance to pesticides and drugs) are full of such adaptive responses.

Sixth and finally, several authors treated population size as constrained by multiple interdependent factors and described the interdependence in system models. System models are large sets of difference equations (deterministic or stochastic), which are usually solved numerically on a computer. System models of human population and other variables have often embodied relationships and assumptions that were neither mechanistically derived nor quantitatively tested.[24]

The first five methods are deterministic and static. They make no allowances for changes in exogenous or endogenous variables or in functional relations among variables. While a probabilistic measure of human carrying capacity has been developed for local populations in the Amazon,[25] no probabilistic approach to global human carrying capacity has been developed. Yet stochastic variability affects local and global human populations through weather, epidemics, accidents, crop diseases and pests, volcanic eruptions, the El Niño Southern Oscillation in the Pacific Ocean, genetic variability in viruses and other microbes, and international financial and political arrangements. Stochastic models of human carrying capacity would make it possible to address questions that deterministic models cannot, such as: conditional on all the assumptions that go into any measure of human carrying capacity, what level of population could be maintained ninety-five years in one hundred in spite of anticipated variability?[26]

Some have urged that individual nations or regions estimate their human carrying capacity separately.[27] While specific resources such as mineral deposits can be defined region by region, the knowledge, energy, and technology required to exploit local resources often depend on other regions; the positive and negative effects of resource development commonly cross national borders. Human carrying capacity cannot be defined for a nation independently of other regions if that nation trades with others and shares the global resources of the atmosphere, oceans, climate, and biodiversity.

Some ecologists and others claim that the ecological concept of carrying capacity provides special insight into the question of how many people the earth can support. In basic and applied ecology, the capacity for carrying nonhuman species has been defined in at least nine significantly different ways, none adequate for humans.[28] Human carrying capacity depends both on natural constraints, which are not fully understood, and on individual and collective choices. How many people the earth can support depends in part on how many will wear cotton and how many polyester; on how many will eat meat and how many bean sprouts; on how many will want parks and how many will want parking lots. These choices will change in time and so will the number of people the earth can support.

The deceptively simple question "How many people can the earth support?" hides a host of thorny issues: How many people with what fashions, tastes, and values? How many people at what average level of material well-being? With what distribution of material well-being? With what technology? With what domestic and international political institutions? With what domestic and international economic and demographic arrangements? In what physical, chemical, and biological environments? With what variability or stability? With what risk or robustness? What standards of personal liberty will people choose?

How many people for how long? Human carrying capacity depends strongly on the time horizon people choose for planning. The population that the earth can support at a given level of well-being for twenty years may differ substantially from the population that can be supported for one hundred or one thousand years.

Mathematical Cartoons

If a current global human carrying capacity could be defined as a statistical indicator, there would be no reason to expect that indicator to be static. In 1798, Thomas Robert Malthus (1766–1834) described a dynamic relation between human population size and human carrying capacity: "The happiness of a country does not depend, absolutely, upon its poverty or its riches, upon its youth or its age, upon its being thinly or fully inhabited, but upon the rapidity with which it is increasing, upon the degree in which the yearly increase of food approaches to the yearly increase of an unrestricted population."[29] Malthus opposed the optimism of the Marquis de Condorcet (1743–94), who saw the human mind as capable of removing all obstacles to human progress. Malthus predicted wrongly that the population growth rate would always promptly win a race against the rate of growth of food. Malthus has been wrong for nearly two centuries because he did not foresee how much people can expand the human carrying capacity of earth, including but not limited to food production. To examine whether Malthus will continue to be wrong, economists, demographers, and system analysts have constructed models in which population growth drives technological change, which permits further population growth.[30]

These models illuminate the earth's human carrying capacity. First, the statement that "every human being represents hands to work, and not just another mouth to feed" does not specify the cultural, environmental, and economic resources available to make additional hands productive, and therefore does not specify by how much the additional hands can increase (or decrease) human carrying capacity; yet the quantitative relation between an increment in population and an increment in carrying capacity is crucial to the future trajectory of both the population and the carrying capacity. Second, the historical record of faster-than-exponential population growth, accompanied by an immense improvement in average well-being, is logically consistent with many alternative futures, including a continued expansion of population and carrying capacity, or a sigmoidal tapering off of the growth in population size and carrying capacity, or oscillations (damped or periodic), or chaotic fluctuations, or overshoot and collapse. Third, to believe that no ceiling to population size or carrying capacity is imminent entails believing that nothing in the near future will stop people from increasing the earth's ability to satisfy their wants by more than, or at least as much as, they consume. The models focus attention on, and provide a framework in which to interpret, quantitative empirical studies of the relation between rapid population growth and changing human carrying capacity.

Issues for the Future

Three valuable approaches have been advocated to ease future trade-offs among population, economic well-being, environmental quality, and cultural values. Each of these approaches is probably necessary, but is not sufficient by itself, to alleviate the economic, environmental, and cultural problems described above.

The "bigger pie" school says: develop more technology.[31] The "fewer forks" school says: slow or stop population growth.[32] In September 1994 at the UN population conference in Cairo, several approaches to slowing population growth by lowering fertility were advocated and disputed. They included promoting modern contraceptives; promoting economic development; improving the survival of infants and children;

improving the status of women; educating men; and various combinations. Unfortunately, there appears to be no believable information to show which approach will lower a country's fertility rate the most, now or a decade from now, per dollar spent. In some developing countries such as Indonesia, family planning programs interact with educational, cultural and, economic improvements to lower fertility by more than the sum of their inferred separate effects.[33] Some unanswered questions are: how soon will global fertility fall? by what means? at whose expense?

The "better manners" school says: improve the terms under which people interact (for example, by defining property rights to open-access resources; by removing economic irrationalities; and by improving governance).[34] When individuals use the environment as a source or a sink and when they have additional children, their actions have consequences for others. Economists call "externalities" the consequences that fall on people who are not directly involved in a particular action. That individuals neglect negative externalities when they use the environment has been called "the tragedy of the commons"[35]; that individuals neglect negative externalities when they have children has been called "the second tragedy of the commons."[36] The balance of positive and negative externalities in private decisions about fertility and use of the environment depends on circumstances. The balance is most fiercely debated when persuasive scientific evidence is least available. Whatever the balance, the neglect by individuals of the negative externalities of childbearing biases fertility upward compared to the level of aggregate fertility that those same individuals would be likely to choose if they could act in concert or if there were a market in the externalities of childbearing. Voluntary social action could change the incentives to which individuals respond in their choices concerning childbearing and use of the environment.

Notes

1. L. R. Brown and H. Kane, *Full House: Reassessing the Earth's Population Carrying Capacity* (New York: W.W. Norton, 1994); F. Duchin and G. Lange, *The Future of the Environment: Ecological Economics and Technological Change* (New York: Oxford University Press, 1994); N. Myers and J. L. Simon, *Scarcity or Abundance? A Debate on the Environment* (New York: W.W. Norton, 1994).

2. M. Livi-Bacci, *A Concise History of World Population*, trans. Carl Ipsen (Oxford, England: Blackwell, 1992). Estimates of global population size at A.D. 1 vary from 133 million (E. S. Deevey, Jr., *Scientific American* 203,

no. 9 [September 1960]: 195–204) to 330 million (J. Durand, *Population and Development Review* 3, no. 3 [1977]: 253–96).

3. *1998 World Population Data Sheet* (Washington, D.C.: Population Reference Bureau, 1998).

4. P. Demeny in *Resources, Environment, and Population: Present Knowledge, Future Options* K. Davis and M. S. Bernstam, eds. (New York: Oxford University Press, 1991), p. 416, gives a grim list of such environmental problems: "loss of topsoil, desertification, deforestation, toxic poisoning of drinking water, oceanic pollution, shrinking wetlands, overgrazing, species loss, shortage of firewood, exhaustion of oil reserves and of various mineral resources, siltation in rivers and estuaries, encroachment of human habitat on arable land, dropping water tables, erosion of the ozone layer, loss of wilderness areas, global warming, rising sea levels, nuclear wastes, acid rain."

5. J. Lederberg, *Journal of the American Medical Association* 260, no. 5 (1988): 684–85; S. S. Morse, ed., *Emerging Viruses* (New York: Oxford University Press, 1993); R. M. Anderson and R. M. May, *Infectious Diseases of Humans: Dynamics and Control* (Oxford, England: Oxford University Press, 1991), chap. 23.

6. In 1960, the richest countries, with 20 percent of world population, earned 70.2 percent of global income, while the poorest countries, with 20 percent of world population, earned 2.3 percent of global income. Thus, the ratio of income per person between the top fifth and the bottom fifth was 31:1. In 1970, that ratio was 32:1; in 1980, 45:1; in 1991, 61:1. In U.S. dollars, the absolute gap between the top fifth and the bottom fifth rose from $1,864 in 1960 to $15,149 in 1989 (United Nations Development Programme, *Human Development Report 1992* [New York: Oxford University Press, 1992], p. 36, *Human Development Report 1994*, p. 63).

7. P. Demeny, *Population and Development, International Conference on Population and Development 1994* (Liège, Belgium: International Union for the Scientific Study of Population, 1994); these numbers are based on World Bank estimates.

8. From 1950 to 1995, the world's urban population increased more than 3.5-fold, from 0.74 billion to 2.6 billion, and from 29 percent to 45 percent of the total population (United Nations, *World Urbanization Prospects: The 1992 Revision,* ST/ESA/SER.A/136 [New York: United Nations, 1993), pp. 74–75, 82–83.

9. *1998 World Population Data Sheet; World Population 1994* (New York: United Nations, 1994).

10. United Nations, Department of International Economic and Social Affairs, *Long-Range World Population Projections: Two Centuries of Population Growth 1950–2150,* ST/ESA/SER.A/125 (New York: United Nations, 1992).

11. Systematic retrospective analyses of past population projections indicate that more confidence has been attached to projections than was justified

by their eventual predictive accuracy; M. A. Stoto, *Journal of the American Statistical Association* 78 (1983): 13–20; N. Keyfitz, *Population Change and Social Policy* (Cambridge, Mass.: Abt Books, 1982), chap. 13.

12. A. van Leeuwenhoek, *Collected Letters* (Amsterdam: Swets and Zeitlinger, 1948), Letter 43, April 25, 1679, vol. 3, pp. 4–35. Leeuwenhoek multiplied his estimate of the population of Holland (1 million people) by his estimate of the ratio of the earth's habitable land area to Holland's area (13,385).

13. J. E. Cohen, *How Many People Can the Earth Support?* (New York: W.W. Norton, 1995).

14. V. Smil, *Population and Development Review* 20 (June 1994): 255–92, estimated 10–11 billion; D. Pimentel, R. Harman, M. Pacenza, J. Pecarsky, and M. Pimentel, *Population and Environment* 15, no. 5 (May 1994): 347–69, estimated < 3 billion; P. E. Waggoner "How Much Land Can Ten Billion People Spare for Nature?" in *Task Force Report No. 121* (Ames, Iowa: Council for Agricultural Science and Technology, February 1994) estimated at least 10 billion; Brown and Kane, *Full House*, p. 202, estimated that a projected world grain harvest of 2.1 billion tons in 2030 could feed 2.5 billion people at the U.S. consumption level of 800 kilograms per year per person, or just over 10 billion people at the Indian level of consumption of 200 kilograms per year per person; Wetenschappelijke Raad voor het Regeringsbeleid [Scientific Council for the Dutch Government], *Duurzame risicos: een blijvend gegeven* [*Sustainable risks: an enduring given*] (Den Haag, The Netherlands: Sdu Uitgeverij Plantijnstraat, 1994), p. 9, estimated 11–44 billion people, depending on the scenario.

15. For example, R. Pearl and L. J. Reed, in R. Pearl, ed., *Studies in Human Biology* (Baltimore: Williams and Wilkins, 1924), chap. 25, p. 632, fitted a logistic curve to past world population sizes and confidently estimated a maximum world population of 2 billion. The world's population passed 2 billion around 1930. Undeterred, R. Pearl and S. Gould, *Human Biology* 8, no. 3 (1936): 399–419, again used the logistic curve to project 2.645 billion people as an ultimate limit to be nearly approached by the end of the twenty-first century. That population size was surpassed before 1955. On a logarithmic scale of population, the logistic curve is concave, while the observed trajectory of global population size was convex until about 1970. The failures of Pearl's logistic projections and the usefulness of Alfred J. Lotka's theory of population growth and age-composition (*Théorie analytique des associations biologiques, II Analyse démographique avec application particulière à l'espèce humaine* [Paris, Hermann: 1939]) led demographers to abandon studying the absolute size of populations in favor of studying population structure and change. Since World War II, estimates of the earth's human carrying capacity have been published almost exclusively by nondemographers. Demography, like economics, still lacks a working theory of scale. In another example of curve fitting, A. L. Austin and J. W. Brewer, *Technological Forecasting and Social Change* 3, no. 1

(1971): 23–49, modified the logistic curve to allow for faster than exponential growth followed by leveling off; they fitted their curve to past global population sizes and predicted an asymptote around 50 billion people.

16. H. Brown, *The Challenge of Man's Future: An Inquiry Concerning the Condition of Man During the Years That Lie Ahead* (New York: Viking, 1954); H. Brown, J. Bonner, and J. Weir, *The Next Hundred Years: Man's Natural and Technological Resources* (New York: Viking, 1957); C. Clark in *Nature* 181 (May 3, 1958): 1235–36, reprinted in W. Y. Davis, ed., *Readings in Human Population Ecology* (Englewood Cliffs, N.J.: Prentice-Hall, 1971), pp. 101–06; M. Cépède, F. Houtart, and L. Grond, *Population and Food* (New York: Sheed & Ward, 1964); W. R. Schmitt in *Annals of the New York Academy of Science* 118, no. 17 (1965): 645–718; H. Lieth in *Human Ecology* 1, no. 4 (1973): 303–32; K. Blaxter, *People, Food and Resources* (Cambridge, England: Cambridge University Press, 1986); P. Buringh, H. D. J. van Heemst, and G. J. Staring, *Computation of the Absolute Maximum Food Production of the World* (Wageningen, The Netherlands: Agricultural Press, 1975); P. Buringh and H. D. J. van Heemst, *An Estimation of World Food Production Based on Labour-Oriented Agriculture* (Wageningen, The Netherlands: Agricultural Press, 1977); G. M. Higgins, A. H. L. Kassam, L. Naiken, G. Fischer, and M. M. Shah, *Potential Population Supporting Capacities of Lands in the Developing World: Technical Report of Project INT/75/P13*, "Land Resources for Populations of the Future," FPA/INT/513 (Rome: Food and Agricultural Organization of the United Nations, 1983); Robert S. Chen et al., eds., *The Hunger Report: 1990* (HR-90-1, Alan Shawn Feinstein World Hunger Program, Brown University, June 1990); S. R. Millman et al., *The Hunger Report: Update 1991* (HR-91-1, Alan Shawn Feinstein World Hunger Program, Brown University, April 1991); Brown and Kane, *Full House*, p. 31.

It is remarkable that food continues to be viewed as a limiting constraint on population size even though, globally, the countries with the lowest fertility and the lowest population growth rates are among those where food is most abundant (J. Mayer, *Daedalus* 93:3 (1964): 830–44.

17. A. Penck, *Zeitschrift für Geopolitik* 2 (1925): 330–48; *Sitzungsberichte der Preussischen Akademie der Wissenschaften* 22 (1924): 242–57. The formula was used but not stated explicitly in 1917 by G. H. Knibbs, *The Mathematical Theory of Population, of Its Character and Fluctuations, and of the Factors Which Influence Them*, Appendix A (Melbourne: Census of the Commonwealth of Australia, Minister of State for Home and Territories; McCarron, Bird & Co.), vol. 1, p. 455.

18. In 1972, domestic animals were fed 41 percent of all grain consumed; in 1992, 37 percent (Brown and Kane) *Full House*, p. 67; World Resources Institute, *World Resources 1994–95* [New York: Oxford University Press, 1994], p. 296).

19. S. R. Eyre, *The Real Wealth of Nations* (New York: St. Martin's Press; London: Edward Arnold, 1978).

20. A. H. Westing, *Environmental Conservation* 8, no. 3 (1981): 177–83.
21. J. F. von Liebig, *Principles of Agricultural Chemistry* (New York: John Wiley, 1855); German edition: *Die Grundsätze der Agriculturchemie* (Braunschweig: F. Vieweg und Sohn). Also see D. L. DeAngelis, *Dynamics of Nutrient Cycling and Food Webs* (London: Chapman and Hall, 1992), pp. 38–45, 228.
22. Liebig's law extends to any number of independent constraints. When population on the left side of the formula is replaced by production, the formula is known in economic theory as the Walras-Leontief-Harrod-Domar production function.
23. V. Smil, *Population and Development Review* 17, no. 4 (1991): 569–601, pp. 586, 597, reported that in the 1980s, nitrogen applied in Zhejiang and Shandong provinces of China increased rice yields by amounts that were only 50–60 percent as large as the increments from an additional kg/ha applied in the 1960s.
24. Examples of system models are: J. W. Forrester, *World Dynamics* (Cambridge, MA: Wright-Allen Press, 1971); D. Meadows, D. L. Meadows, J. Randers, and William W. Behrens III, *The Limits to Growth* (New York: New American Library, 1972); M. Mesarovic and E. Pestel, *Mankind at the Turning Point* (New York: E. P. Dutton and Reader's Digest Press, 1974); P. House and E. R. Williams, *The Carrying Capacity of a Nation: Growth and the Quality of Life* (Lexington, Mass.: D. C. Heath, 1975); J. Gever, R. Kaufmann, D. Skole, and C. Vörösmarty, *Beyond Oil: The Threat to Food and Fuel in the Coming Decades,* 3rd ed. (Niwot: University Press of Colorado, 1991); A. J. Gilbert and L. C. Braat, eds., *Modelling for Population and Sustainable Development* (London: Routledge, 1991); W. Lutz, ed., *Population-Development-Environment: Understanding Their Interactions in Mauritius* (Heidelberg: Springer-Verlag, 1994). Critiques of systems models are: C. Kaysen in *Foreign Affairs* 50, no. 4 (July 1972): 660–68; H. S. D. Cole, C. Freeman, M. Jahoda, and K. L. R. Pavitt, eds., *Models of Doom: A Critique* of *The Limits to Growth,* with a reply by the authors of *The Limits to Growth* (New York: Universe Books, 1973); W. D. Nordhaus, *Economic Journal* 83, 332 (December 1973): 1156–83; E. van de Walle, in *Science* 189 (1975): 1077–78; D. Berlinski, *On Systems Analysis* (Cambridge, Mass.: MIT Press, 1976); P. R. Ehrlich, A. H. Ehrlich, and J. P. Holdren, *Ecoscience: Population, Resources, Environment* (San Francisco: W.H. Freeman, 1977), pp. 730–33.
25. P. M. Fearnside, *Human Carrying Capacity of the Brazilian Rainforest* (New York: Columbia University Press, 1986).
26. In many regions, the average amount of fresh water available annually is more than twice the amount of water that can be counted on ninety-five years in one hundred; (P.P. Rogers in *The Global Possible: Resources, Development, and the New Century,* R. Repetto, ed. (New Haven: Yale University Press, 1985), p. 294.
27. Paragraph 5.23 of *Agenda 21,* final document of the United Nations Con-

ference on Environment and Development, held in Rio de Janeiro in June 1992 (Rio Earth Summit).

28. B. Zaba and I. Scoones in *Environment and Population Change*, B. Zaba and J. Clarke, eds. (Liege, Belgium: Ordina Editions, 1994), pp. 197–219; H. R. Pulliam and N. M. Haddad, *Bulletin of the Ecological Society of America* 75 (1994): 141–57; J. E. Cohen, *How Many People Can the Earth Support?* chap. 12.

29. T. R. Malthus, *An Essay on the Principle of Population*, complete 1st ed. (1798) and partial 7th ed. (1872) reprinted in Gertrude Himmelfarb, ed., *On Population* (New York: Modern Library, 1960), chap. 7, p. 51. P. Demeny, in *Population and Resources in Western Intellectual Traditions*, M. S. Teitelbaum and J. M. Winter, eds. (New York: Population Council, 1989), p. 232, generalized Malthus's view to incorporate all aspects of economic output, not just food: "Posed in the simplest terms, the economics of population reduces to a race between two rates of growth: that of population and that of economic output."

30. F. L. Pryor and S. B. Maurer in *Journal Development Economics* 10 (1982): 325–53; R. D. Lee in *The State of Population Theory: Forward from Malthus*, D. Coleman and R. Schofield, eds. (Oxford: Basil Blackwell, 1986), pp. 96–130; R. D. Lee in *Mathematical Population Studies* 1, no. 3 (1988): 265–88; R. D. Lee, *Population* (Paris) 47, no. 6 (1992): 1533–54; R. D. Lee in *Explorations in Economic History* 30 (1993): 1–30; M. Kremer in *Quarterly Journal of Economics* 108, no. 3 (1993): 681–716.

31. J. H. Ausubel in *The Sciences* (New York Academy of Sciences) 33, no. 6 (1993): 14–19; P. E. Waggoner, "How Much Land Can Ten Billion People Spare for Nature?"

32. J. Bongaarts in *Science* 263 (1994): 771–76.

33. P. J. Gertler and J. W. Molyneaux in *Demography* 31, no. 1 (1994): 33–63.

34. R. Repetto, ed., *The Global Possible: Resources, Development, and the New Century* (New Haven, Conn.: Yale University Press, 1985); D. W. Pearce and J. J. Warford, *World Without End: Economics, Environment, and Sustainable Development* (New York: Oxford University Press, 1993).

35. G. Hardin in *Science* 162 (1968): 1243–48.

36. R. D. Lee in Kingsley Davis and Mikhail S. Bernstam, eds., *Resources, Environment, and Population: Present Knowledge, Future Options* (New York: Oxford University Press, 1991), pp. 315–22. Earlier discussions of the negative externalities of childbearing include H. F. Dorn in *Science* 135, no. 3500 (1962): 283–90; P. E. Sorenson in N. Hinrichs, ed., *Population, Environment and People* (New York: McGraw-Hill, 1971), pp. 113–21; P. Dasgupta and R. J. Willis in D. G. Johnson and R. D. Lee, eds., *Population Growth and Economic Development: Issues and Evidence* (Madison: University of Wisconsin Press, 1987), pp. 631–702.

37. C. McEvedy and R. Jones, *Atlas of World Population History* (New York: Viking Penguin, 1978); J.-N. Biraben in *Population* (Paris) 34, no. 1 (1979): 13–25. See also Durand, *Population and Development Review*.

38. C. M. Cipolla, *Economic History of World Population,* 6th ed. (Harmondsworth, England: Penguin, 1974), pp. 56–9; World Resources Institute, *World Resources 1994–95* (New York: Oxford University Press, 1994), pp. 332–34.

39. Data are from note 13, Cohen, *How Many People.* The estimate by J. H. Fremlin in *New Scientist* 24, no. 415 (October 29, 1964): 285–87, would be off the scale and is omitted.

Revisiting Carrying Capacity:
Area-Based Indicators of Sustainability

William E. Rees*

Conventional wisdom suggests that because of technology and trade, human carrying capacity is infinitely expandable and therefore virtually irrelevant to demography and development planning. By contrast, this essay argues that ecological carrying capacity is the primary basis for demographic accounting. A fundamental question for ecological economics is whether remaining stocks of natural capital are adequate to sustain the anticipated load of the human economy into the next century. Since mainstream (neoclassical) models are blind to ecological structure and function, they cannot even properly address this question. The present essay therefore assesses the capital stocks, physical flows,

*William E. Rees is a professor at and director of the School of Community and Regional Planning, University of British Columbia. This essay was originally published in shorter form in *Population and Environment: A Journal of Interdisciplinary Studies* 17, no. 3 (January 1996): 195–215. Copyright © 1996 Human Sciences Press, Inc. The original essay was revised from a presentation of the same title to the International Workshop on Evaluation Criteria for a Sustainable Economy, Institut für Verfahrenstechnick, Technische Universität Graz, Graz, Austria, April 6–7, 1994. It has been expanded for this book using material drawn from "Reducing the Ecological Footprints of Consumption," the author's contribution to *Toward the Goal of a Sustainable Society: Policy Measures for Changing Consumption Patterns*, proceedings of a workshop sponsored by the Korean Environmental Technology Research Institute, Seoul, Korea, August 30–September 1, 1995. The author's and his students' work on ecological footprinting is supported by a Canadian Tri-Council EcoResearch Grant to the University of British Columbia, in which the author is a co-investigator.

and corresponding ecosystems areas required to support the economy using "ecological footprint" analysis. This approach shows that most so-called advanced countries are running massive unaccounted ecological deficits with the rest of the planet. Since not all countries can be net importers of carrying capacity, the material standards of the wealthy cannot be extended sustainably to even the present world population using prevailing technology. In this light, sustainability may well depend on such measures as greater emphasis on equity in international relationships, significant adjustments to prevailing terms of trade, increasing regional self-reliance, and policies to stimulate a massive increase in the material and energy efficiency of economic activity.

Why Carrying Capacity?

According to Garrett Hardin (1991), "Carrying capacity is the fundamental basis for demographic accounting." On the other hand, conventional economists and planners generally ignore or dismiss the concept when applied to human beings. Their vision of the human economy is one in which "the factors of production are infinitely substitutable for one another" and in which "using any resource more intensely guarantees an increase in output" (Kirchner et al., 1985). As Daly (1986) observes, this vision assumes a world "in which carrying capacity is infinitely expandable" (and therefore irrelevant). Clearly, there is great division over the value of carrying-capacity concepts in the sustainability debate.

This essay sides solidly with Hardin. I start from the premise that despite our increasing technological sophistication, humankind remains in a state of "obligate dependence" on the productivity and life-support services of the ecosphere (Rees 1990). Thus, from an ecological perspective, adequate land and associated productive natural capital are fundamental to the prospects for continued civilized existence on earth. However, at present, both the human population and average consumption are increasing, while the total area of productive land and stocks of natural capital are fixed or in decline. These opposing trends demand a revival of carrying-capacity analysis in sustainable development planning. The complete rationale is as follows:

Definitions: Carrying Capacity and Human Load

An environment's carrying capacity is its maximum persistently supportable load (Catton, 1986). For purposes of game and range management, carrying capacity is usually defined as the maximum population of a given species that can be supported indefinitely in a defined habitat without permanently impairing the productivity of that habitat.

However, because of our seeming ability to increase human carrying capacity by eliminating competing species, by importing locally scarce resources, and through technology, this definition seems irrelevant to humans. Indeed, trade and technology are often cited as reasons for rejecting the concept of human carrying capacity out of hand.[1]

This is an ironic error—shrinking carrying capacity may soon become the single most important issue confronting humanity. The reason for this becomes clearer if we define carrying capacity not as a maximum population but rather as the maximum "load" that can safely be imposed on the environment by people. Human load is a function not only of population but also of per capita consumption, and the latter is increasing even more rapidly than the former, due (ironically) to expanding trade and technology. As Catton (1986) observes: "The world is being required to accommodate not just more people, but effectively 'larger' people. . . ." For example, in 1790 the estimated average daily energy consumption by Americans was 11,000 kcal. By 1980, this had increased almost twentyfold to 210,000 kcal/day (Catton 1986). As a result of such trends, load pressure relative to carrying capacity is rising much faster than is implied by mere population increases.

The Ecological Argument

Despite our technological, economic, and cultural achievements, achieving sustainability requires that we understand human beings as ecological entities. Indeed, from a functional perspective, the relationship of humankind to the rest of the ecosphere is similar to that of millions of other species with which we share the planet. We depend for both basic needs and the production of artifacts on energy and material resources extracted from nature, and all this energy-matter is eventually returned in degraded form to the ecosphere as waste. The major material difference between humans and other species is that in addition to our biological metabolism, the human enterprise is characterized by an industrial metabolism. In ecological terms, all our toys and tools (the "capital" of economists) are "the exosomatic equivalent of organs" (Sterrer 1993) and, like bodily organs, require continuous flows of energy and material to and from "the environment" for their production and operation. It follows that in a finite world:

• Economic assessments of the human condition should be based on, or at least informed by, ecological and biophysical analyses.
• The appropriate ecological analyses focus on the flows of available energy-matter (essergy) particularly from primary producers—green plants and other photosynthesizers—to sequential levels of consumer

organisms in ecosystems (specifically, humans and their economies) and on the return flows of degraded energy and material (wastes) back to the ecosystem.

This approach shows that humankind, through the industrial economy, has become the dominant consumer in most of the earth's major ecosystems. We currently "appropriate" 40 percent of the net product of terrestrial photosynthesis (Vitousek et al., 1986) and 25–35 percent of coastal shelf primary production (Pauly and Christensen, 1995), and these may be unsustainable proportions.[2] At the same time, some global waste sinks seem full to overflowing.

A fundamental question for *ecological* economics, therefore, is whether the physical output of remaining species populations, ecosystems, and related biophysical processes (i.e., critical self-producing natural capital stocks—see box 1), and the waste assimilation capacity of

---

Box 1. On Natural Capital

Natural capital refers to "a stock [of natural assets] that yields a flow of valuable goods and services into the future." For example, a forest or a fish stock can provide a flow or harvest that is potentially sustainable year after year. The stock that produces this flow is "natural capital" and the sustainable flow is "natural income." Natural capital also provides such services as waste assimilation, erosion and flood control, and protection from ultra-violet radiation (the ozone layer is a form of natural capital). These life-support services are also counted as natural income. Since the flow of services from ecosystems often requires that they function as intact systems, the structure and diversity of the system may be an important component of natural capital.

There are three broad classes of natural capital: *Renewable* natural capital, such as living species and ecosystems, is self-producing and self-maintaining using solar energy and photosynthesis. These forms can yield marketable goods such as wood fiber, but may also provide unaccounted essential services when left in place (e.g., climate regulation). *Replenishable* natural capital, such as groundwater and the ozone layer, is nonliving but is also often dependent on the solar "engine" for renewal. Finally, *nonrenewable* natural capital, such as fossil fuel and minerals, is analogous to inventories—any use implies liquidating part of the stock.

This essay takes the position that since adequate stocks of self-producing and replenishable natural capital are essential for life support (and are generally nonsubstitutable), these forms are more important to sustainability than are nonrenewable forms.

Source: Rees (1995), liberally adapted from Costanza and Daly (1992).

the ecosphere, are adequate to sustain the anticipated load of the human economy into the next century while simultaneously maintaining the general life-support functions of the ecosphere. This "fundamental question" is at the heart of ecological carrying capacity but is virtually ignored by mainstream analyses.

Second Law Arguments

A related rationale for revisiting carrying capacity flows from consideration of the second law of thermodynamics. In particular, modern formulations of the second law suggest that all highly ordered systems develop and grow (increase their internal order) "at the expense of increasing disorder at higher levels in the systems hierarchy" (Schneider and Kay 1994). In other words, complex dynamic systems remain in a nonequilibrium state through the continuous dissipation of available energy and material (essergy) extracted from their host environments. They require a constant input of energy-matter to maintain their internal order in the face of spontaneous entropic decay. Such self-organizing nonequilibrium systems are therefore called dissipative structures. This extension of the second law is critical to human carrying capacity. Consider that:

- The human economy is one such highly ordered, dynamic, far-from-equilibrium dissipative structure. At the same time:
- The economy is an open, growing, subsystem of a materially closed, nongrowing ecosphere (Daly, 1992) and is therefore dependent on the formation of essergy in the ecosphere for its growth and development.[3]

This relationship implies that beyond a certain point, the continuous growth of the economy can be purchased only at the expense of increasing disorder or entropy in the ecosphere. This is the point at which consumption by the economy exceeds natural income and would be manifested through the continuous depletion of natural capital—reduced biodiversity, air/water/land pollution, deforestation, atmospheric change, etc. In other words, the empirical evidence suggests that the aggregate human load already exceeds, and is steadily eroding, the very carrying capacity upon which continued human existence depends. Ultimately, this poses the threat of unpredictable ecosystems restructuring (e.g., erratic climate change) leading to resource shortages, increased local strife, and the heightened threat of ecologically induced geopolitical instability.

In this light, the behavior of complex systems and the role of the economy in the global thermodynamic hierarchy should be seen as fundamental to sustainability, yet both concepts are alien to the dominant development-oriented institutions in the world today.

Why Economics Cannot Cope

Part of the reason for this perceptual gulf is that many of the questions raised by ecological and thermodynamic considerations are invisible to mainstream approaches. Economic analysis is based on the circular flow of exchange value (money flows) through the economy, not on physical flows and transformations. Prevailing economic models of growth and sustainability thus "lack any representation of the materials, energy sources, physical structures, and time-dependent processes basic to an ecological approach" (Christensen, 1991). Thus, while the second law is arguably the ultimate governor of economic activity, standard models do not recognize the unidirectional and thermodynamically irreversible flux of available energy and matter upon which the economy depends (figure 1). Similarly, conventional approaches to con-

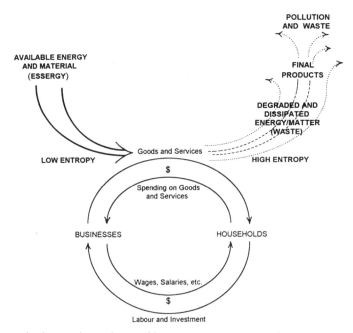

Figure 1. The linear throughput of low-entropy energy and matter (upper part of diagram) sustains the economy and drives the circular flows of exchange value (lower part of diagram) yet is invisible to conventional economic analysis. *Source:* Rees (1995).

Box 2. The Blind Spot in Conventional Analysis

Mainstream economics approaches the issue of adequate capital stocks through monetary analysis. However, money and prices are excessively abstracted from the material wealth they are supposed to represent. For example: Where there are markets for ecologically significant "goods and services," prices do not reflect the size of the corresponding natural capital stocks, whether there are critical minimal levels below which stocks can no longer replenish themselves (the real measure of biophysical scarcity), the functional roles of such stocks in relevant ecosystems, or their ultimate value in sustaining life. Meanwhile, many ecological goods and most life-support services remain unpriced and therefore are not subject to market signals or related behavioral change of *any* kind. (The ozone layer is a case in point.)

Unfortunately, current efforts to "internalize the externalities," "get the prices right," and otherwise commodify the environment suffer from major data gaps, the functional transparency of natural processes (we don't know they're valuable until they're gone), and other theoretical problems that often render futile attempts to quantify, let alone price, many critical ecological goods and services (Vatn and Bromley 1993).

In short, standard monetary analyses are blind to ecological structure and function and are therefore incapable of indicating either ecologically meaningful scarcity or incipient systems destabilization.

---

servation and sustainability focus mainly on the money values of marketable resource commodities (e.g., timber) and are insensitive to the intangible (but ultimately more valuable) nonmarket ecological functions of the natural capital that produces them (e.g., the forest ecosystem). Box 2 summarizes this problem.

In this light, economists' lack of concern about carrying capacity would seem to derive, in large part, from conceptual weaknesses in their analytic models. The necessary conditions for ecological sustainability can better be defined through the analysis of physical stocks and flows interpreted in light of appropriate ecological and complex systems theory.

Technology and Trade: No Boon to Carrying Capacity

As previously noted, conventional analysts often argue that trade and technology expand ecological carrying capacity. This is a misconception. Even in the best of circumstances, technological innovation increases not carrying capacity per se but only the efficiency of resource use. In theory, shifting to more energy- and material-efficient technologies should enable a defined environment to support a given population at a higher material standard, or a higher population at the same material standard, thereby seeming to increase carrying capacity. However, in either case, the best we could hope for in an increasingly open global economy would be to keep total human load constant in the vicinity of carrying capacity—the latter would still ultimately be limiting.

In practice, we have not done even this well—the steady gains in efficiency throughout the postwar period have been accompanied by steadily increasing per capita and aggregate consumption. It seems that efficiency gains may actually work against conservation through the price and income effects of technological savings.

As Saunders (1992) notes, this counterintuitive hypothesis has been the focus of considerable controversy. He tested it using neoclassical growth theory and found that energy-efficiency gains might well increase aggregate energy consumption by making energy cheaper and by stimulating economic growth, which further "pulls up" energy use. How might this work? If a firm saves money by switching to more energy- and material-efficient manufacturing processes, it will be able to raise wages, increase dividends, or lower prices, which can lead to increased net consumption by workers, shareholders, or consumers respectively. Similarly, technology-induced money savings by individuals are usually redirected to alternative forms of consumption, canceling some or all of the initial potential benefit to the environment (Hannon, 1975). These behavioral responses to changes in prices and income are referred to as the "rebound effect" by economists (Jaccard, 1991). To the extent that such mechanisms contribute to increased aggregate material consumption and accelerated stock depletion, they indirectly *reduce* carrying capacity.[4]

More generally, however, technology can directly reduce carrying capacity while creating the illusion of increasing it. We often use technology to increase the short-term energy and material flux through exploited ecosystems. This seems to enhance systems productivity while actually permanently eroding the resource base. For example, the effectiveness of electronic fish-finding devices and high-tech catching technology has overwhelmed the reproductive capacity of fish stocks. Energy-subsidized intensive agriculture may be more productive than low-input practices in the short term, but it also increases the rate of soil and water depletion. The net effect is to create unsustainable dependencies on enhanced material flows (the technologies involved are often based on nonrenewable resources) while reducing long-term carrying capacity.

The carrying-capacity gains from trade are also illusory. While commodity trade may release a local population from carrying-capacity constraints in its own home territory, this merely displaces some fraction of that population's environmental load to distant export regions. In effect, local populations import others' "surplus" carrying capacity. The resultant increase in population and resource use in import regions increases the aggregate load of humanity on the ecosphere, but there is *no net gain* in carrying capacity since trade reduces the load-bearing capacity of the export regions. Indeed, like technology, trade may even result in reduced global carrying capacity if access to cheap imports (e.g., food) lowers the incentive for people to conserve their own local natural capital stocks (e.g., agricultural land) and leads to the accelerated depletion of natural capital in distant export regions.

These comments are not to be taken as arguments against technology or trade per se. Rather, the point is to emphasize that conventional assumptions about both should be carefully reexamined in light of carrying-capacity considerations and that certain conditions must be satisfied before either can contribute to ecological sustainability.

Appropriated Carrying Capacity and Ecological Footprints

We can now redefine human carrying capacity as the maximum rates of resource harvesting and waste generation (the maximum load) that can be sustained indefinitely without progressively impairing the productivity and functional integrity of relevant ecosystems wherever the latter may be located. The size of the corresponding population would be a function of technological sophistication and mean per capita material standards (Rees 1988). This definition reminds us that regardless of the state of technology, humankind depends on a variety of ecological goods and services provided by nature and that for sustainability, these must be available in increasing quantities from somewhere on the planet as population and mean per capita resource consumption increase (see also Overby 1985).

Now, as noted earlier, a fundamental question for ecological economics is whether supplies of natural capital will be adequate to meet anticipated demand into the next century. Inverting the standard carrying-capacity ratio suggests a powerful way to address this critical issue. Rather than asking what population a particular region can support sustainably, the carrying-capacity question becomes: How large an area of productive land is needed to sustain a defined population indefinitely, *wherever on earth that land is located* (Rees, 1992; Rees and Wackernagel, 1994; Wackernagel and Rees, 1996)? Since many forms of natural income (resource and service flows) are produced by terrestrial ecosystems and associated water bodies, it should be possible to estimate the area of land/water required to produce sustainably the quantity of any resource or ecological service used by a defined population at a given level of technology. The sum of such calculations for all significant categories of consumption would give us a conservative area-based estimate of the natural capital requirements for that population.

A simple mental exercise serves to illustrate the ecological reality behind this approach. Imagine what would happen to any modern human settlement or urban region, as defined by its political boundaries or the area of built-up land, if it were enclosed in a glass or plastic hemisphere completely closed to material flows. Clearly, the city would cease to function and its inhabitants would perish within a few days. The population and economy contained by the capsule would be cut off from both vital resources and essential waste sinks, leaving it to starve and suffocate at the same time. In other words, the ecosystems contained within our imaginary human terrarium would have insufficient carrying capacity to service the ecological load imposed by the contained population.

This mental model illustrates the simple fact that as a result of high population densities, the enormous increase in per capita energy and material consumption made possible by (and required by) technology, and universally increasing dependencies on trade, *the ecological locations of human settlements no longer coincide with their geographic locations.* Twentieth-century cities and industrial regions are dependent for survival and growth on a vast and increasingly global hinterland of ecologically productive landscapes. It seems that in purely ecological terms, modern settlements have become the human equivalent of cattle feedlots.

Cities necessarily appropriate the ecological output and life-support functions of distant regions all over the world through commercial trade and the natural biogeochemical cycles of energy and material. Indeed, the annual flows of natural income required by any defined population can be called its *appropriated carrying capacity.* Since for every material flow there must be a corresponding land/ecosystem source or sink, the total area of land/water required to sustain these flows on a continuous basis is the true "ecological footprint" of the referent population on the earth. (See box 3 for definitions of these and related indicators.) Calculating its ecological footprint provides a rough

---

Box 3. A Family of Area-Based Sustainability Indicators

- Appropriated carrying capacity: The biophysical resource flows and waste assimilation capacity appropriated per unit time from global totals by a defined economy or population.
- Ecological footprint: The corresponding area of productive land and aquatic ecosystems required to produce the resources used, and to assimilate the wastes produced, by a defined population at a specified material standard of living, wherever on earth that land may be located.
- Personal planetoid: The per capita ecological footprint (EF_p/N).
- Fair earthshare: The amount of ecologically productive land "available" per capita on earth, currently (1995) about 1.5 hectares. A fair seashare (ecologically productive ocean—coastal shelves upwellings and estuaries—divided by total population) is just over 0.5 ha.
- Ecological deficit: The level of resource consumption and waste discharge by a defined economy or population in excess of locally/regionally sustainable natural production and assimilative capacity (also, in spatial terms, the difference between that economy/population's ecological footprint and the geographic area it actually occupies).
- Sustainability gap: A measure of the decrease in consumption (or the increase in material and economic efficiency) required to eliminate the ecological deficit. (Can be applied on a regional or global scale.)

---

measure of the natural capital requirements of any subject population for comparison with available supply.

"Footprinting" the Human Economy

The first step in calculating the ecological footprint of a study population is to estimate the per capita land area appropriated (aa) for the production of each major consumption item (i) We do this by dividing average annual consumption of that item (c, in kg/capital) by its average annual productivity or yield (p, in kg/ha) per hectare:

$$aa_i = c_i/p_i$$

In practice, it is often only possible to estimate average per capita consumption by dividing aggregate consumption by the referent population size. Of course, many consumption items (e.g., clothing and furniture) embody several inputs, and we have found it useful to estimate the areas appropriated by each significant input separately. Ecological footprint calculations are therefore both more complicated and more interesting than appears from the basic concept. So far we have estimated the land requirements to produce twenty-three categories of consumer goods and services (Wackernagel and Rees, 1994).

We then compute the total per capita ecological footprint (ef) by summing all the ecosystem areas appropriated by individual items in the annual shopping basket of consumption goods and services:

$$\text{ef} = \sum_{i=1}^{i=n} \text{aa}_i$$

Thus, the ecological footprint of a study population (EF_p) is the per capita footprint multiplied by population size (N):

$$EF_p = N(\text{ef})$$

We account for direct fossil energy consumption and the energy content of consumption items by estimating the area of carbon-sink forest that would be required to sequester the carbon dioxide emissions associated with burning fossil fuels ([carbon emissions/capita]/[assimilation rate/hectare]), on the assumption that atmospheric stability is central to sustainability. (An alternative is to estimate the area of land required to produce the biomass energy equivalent [ethanol] of fossil energy consumption. This produces a larger energy footprint than the carbon assimilation method.)

Every effort is made to avoid double counting in the case of multiple land uses, and where there are data problems or significant uncertainty, we err on the side of caution. Also, while we define the footprint comprehensively to include the land/water areas required for waste assimilation, our calculations to date do not account for waste emissions other than carbon dioxide. Accounting fully for this ecological function would add considerably to the ecosystem area appropriated by economic activity. Together these factors suggest that our ecological footprint calculations to date are more likely to be underestimates than overestimates.

Data from my home city, Vancouver, British Columbia, Canada, serve to illustrate application of the concept. Vancouver proper has a population (as of 1991) of 472,000 and an area of 114 km² (11,400 hectares). However, the average Canadian requires over a hectare (ha) of crop and grazing land under current land management practices to produce his or her high meat protein diet and about 0.6 ha for wood and paper associated with various other consumption items. In addition, each "occupies" about 0.2 ha of ecologically degraded and built-over (e.g., urban) land. Canadians are also among the world's highest fossil energy consumers, with an annual carbon emission rate of 4.2 tons of carbon (15.4 tons CO_2) per capita (data corrected for carbon content of trade goods). Therefore, at a carbon sequestering rate of 1.8 tons/ha/yr, an additional 2.3 ha of middle-aged north temperate forest would be required as a continuous carbon sink to assimilate the average Canadian's carbon emissions (assuming the need to stabilize atmospheric carbon dioxide levels).

Considering only these data, the terrestrial "personal planetoid" of a typical Vancouverite approaches 4.2 ha, or almost three times his or her "fair earthshare."[5] On this basis, the 472,000 people living in Vancouver require, conservatively, 2.0 million ha of land for their exclusive use to maintain their current consumption patterns (assuming such land is being managed sustainably). That the area of the city is only about 11,400 ha means that the city population appropriates the productive output of a land area nearly 174 times larger than its political area to support its present consumer lifestyles.[6] While this result might seem extraordinary, other researchers have obtained similar results. Folke, Larsson, and Sweitzer (1994) report that the aggregate consumption of wood, paper, fiber, and food (including seafood) by the inhabitants of twenty-nine cities in the Baltic Sea drainage basin appropriates an ecosystem area 200 times larger than the area of the cities themselves. (The latter study does not include energy land.)

Many whole developed countries have a similar overwhelming dependence on external ecoproductivity. The Netherlands (area 33,920 km²) serves to illustrate: We estimate that the people of Holland require a land area more than fourteen to fifteen times larger than their country to support current domestic consumption of food, forest products, and energy (Rees and Wackernagel, 1994). The food footprint alone is more than 100,000 km², based on world average productivities. Indeed, Dutch government data suggest that the Netherlands appropriates 100,000 to 140,000 km² of agricultural land, mostly from the third world, for food production (including value-added food products produced in the Netherlands for export) (RIVM 1991, cited in Meadows, Meadows, and Randers, 1992).[7] This "imported" land is five to seven times the area of Holland's domestic arable land.

It is worth remembering that the Netherlands, like Japan, is often held up as an economic success story and an example for the developing world to follow. Despite small size, few natural resources, and relatively large populations, both Holland and Japan enjoy high material standards and positive current account and trade balances as measured in monetary terms. However, our analysis of physical flows shows that these and most other so-called advanced economies are running massive, unaccounted, ecological deficits with the rest of the planet (table 1). The last two columns in table 1 represent low estimates of these per capita ecological deficits in a selection of developed countries. Even if their land were twice as productive as world averages, many European countries would still run a deficit more than three times larger than domestic natural income.

These data emphasize that all the countries listed, except for Canada, are overpopulated in ecological terms—they could not sustain them-

Table 1. The Ecological Deficits of Industrialized Countries

	Ecologically Productive Land (in hectares)	Population (1995)	Ecologically Productive Land Per Capita (in hectares)	National Ecological Deficit Per Capita	
				(in hectares)	(in % available)
	a	b	c = a/b	d = Footpr. − c	e = d/c
Countries with 2–3-ha footprints			Assuming a 2-hectare footprint		
Japan	30,340,000	125,000,000	0.24	1.76	730%
Korea	8,669,000	45,000,000	0.19	1.81	950%
Countries with 3–4-ha footprints			Assuming a 3-hectare footprint		
Austria	6,740,000	7,900,000	0.85	2.15	250%
Belgium	1,987,000	10,000,000	0.20	2.80	1,400%
Denmark	3,270,000	5,200,000	0.62	2.38	380%
France	45,385,000	57,800,000	0.78	2.22	280%
Germany	27,734,000	81,300,000	0.34	2.66	780%
Netherlands	2,300,000	15,500,000	0.15	2.85	1,900%
Switzerland	3,073,000	7,000,000	0.44	2.56	580%
Countries with 4–5-ha footprints			Assuming 4.3- (Can.) and 5.1-hectare (U.S.) footprint		
Canada	433,000,000	28,500,000	15.19	(10.89)	(250%)
United States	725,643,000	258,000,000	2.81	2.28	80%

Source: Revised from Wackernagel and Rees (1994).
Note: Footprints were estimated from studies by Ingo Neumann of Trier University, Germany, Dieter Zürcher of Infras Consulting, Switzerland, and our own analysis using World Resources Institute (1992) data.

selves at current material standards if forced by changing circumstances to live on their remaining endowments of domestic natural capital. This is hardly a good model for the rest of the world to follow.

Canada (large area, resource rich, small population) is one of the few developed countries that, at least domestically, consumes less than its natural income. However, Canada's natural capital stocks are being depleted by exports of energy, forest, fisheries, agricultural products, etc. In short, Canada's apparent ecological surpluses are being incorporated in part by trade into the ecological footprints—and deficits—of other countries, particularly those of the United States and Japan.

Sustaining Development with Phantom Planets?

Ecological deficits are a measure of the entropic load and resultant "disordering" being imposed on the ecosphere by so-called advanced countries as the unaccounted cost of maintaining and further expanding their wealthy consumer economies. This massive entropic imbalance invokes what might be called the first axiom of ecological footprint analysis: On a finite planet, not all countries or regions can be net importers of carrying capacity. This, in turn, has serious implications for global development trends.

The current objective of international development is to raise the developing world to present first-world material standards. To achieve this objective, the Brundtland Commission argued for "more rapid economic growth in both industrial and developing countries" and suggested that "a five- to ten-fold increase in world industrial output can be anticipated by the time world population stabilizes some time in the next century" (WCED, 1987).

Let us examine this prospect using ecological footprint analysis. If just the present world population of 5.8 billion people were to live at current North American ecological standards (say, 4.5 ha/person), a reasonable first approximation of the total productive land requirement would be 26 billion ha (assuming present technology). However, there are only just over 13 billion ha of land on earth, of which only 8.8 billion are ecologically productive cropland, pasture, or forest (1.5 ha/person). In short, we would need an additional two planet earths to accommodate the increased ecological load of people alive today. If the population were to stabilize at between 10 and 11 billion sometime in the next century, five additional earths would be needed, all else being equal—and this is just to maintain the present rate of ecological decline (Rees and Wackernagel, 1994).

While this may seem to be an astonishing result, empirical evidence suggests that five phantom planets is, in fact, a considerable underestimate (keep in mind that our footprint estimates are conservative). Global and regional-scale ecological change in the form of atmospheric change, ozone depletion, soil loss, groundwater depletion, deforestation, fisheries collapse, loss of biodiversity, and so forth is accelerating. This is direct evidence that aggregate consumption exceeds natural income in certain critical categories and that the carrying capacity of this one earth is being steadily eroded.[8] In short, the ecological footprint of the present world population/economy already exceeds the total productive area (or ecological space) available on earth.

This situation is, of course, largely attributable to consumption by that wealthy quarter of the world's population who uses 75 percent of global resources. The WCED's "five- to ten-fold increase in industrial output" was deemed necessary to address this obvious inequity while accommodating a much larger population. However, since the world is already ecologically full, sustainable growth on this scale using present technology would require five to ten additional planets.

The Factor-10 Economy

There is no getting around the fact that material consumption is at the heart of the sustainability crisis—the aggregate "ecological footprint" of humanity is already larger than the earth. The ecological challenge for sustainability, therefore, is how to accommodate both rising material expectations and a near-doubling of population over the next fifty years while actually *reducing* total throughput.

Barring disaster, most people seem to agree that this can be achieved in two ways: through an absolute reduction in the average material standard of living or through a massive increase in material and energy efficiency (or through some combination of the above). The increasingly ubiquitous cultural values implied by "consumerism" render the first approach politically unfeasible in developed countries, and it would quite justifiably be rejected outright by the impoverished quarter of the world's population living mostly in developing countries. Indeed, many economists insist that global sustainability is achievable only through large increases in the consumption of goods and services in both the poor and rich countries.

The bad news, then, is that growth is seen as the only politically and economically viable means to alleviate poverty and inequity both within countries and between rich and poor countries. The good news is that many advocates of this approach have at last accepted the fact of limits to *material* growth. A consensus is emerging that the needed increase in consumption would be sustainable only if there is a corresponding reduction in the material and energy intensity of goods and services (see Pearce, 1994).

Numerous researchers and organizations are therefore exploring the policy implications of reducing the energy/material throughput of so-called advanced economies. Conscious of the need for growth, particularly in the developing world, they conclude that the material intensity of consumption in industrial countries should be reduced by a factor of up to ten (factor 10) to accommodate it (BCSD, 1993; Elkins and Jacobs, 1994; Young and Sachs, 1994). Since markets do not reflect ecological reality, governments must create the necessary policy incentives to ensure that as consumption rises, the material and energy content of that consumption falls apart. Achieving a factor-10 economy will require major changes in industrial strategy, fiscal and taxation policy, and consumer–corporate relations. However, if managed properly, the net effect of this transformation should be not only less consumption and waste but also more jobs and increased regional self-reliance.

An Alternative Approach: Investing in Social Capital

The anticipated factor-10 efficiency revolution is based on the assumption that economic growth everywhere is essential for sustainability. It also requires unbridled optimism in technological innovation. It is therefore a predictable response from the industrial–scientific or expansionist paradigm that prevails in international development circles today.

The problem is that technological fixes address neither the fundamental cultural values nor the growth ethic that has produced the ecological crisis and that lies at the heart of the mainstream paradigm. Arguably, therefore, by focusing exclusively on potential efficiency gains, policy makers may overlook other effective paths to ecologically sustainable socioeconomic development. One option is to consider the efficacy of investing in social capital. If building up our stocks of social capital can substitute for the perceived need to accumulate manufactured capital, then *large reductions in society's ecological footprint may be possible even without technological efficiency gains.*

Welfare and Income

There are at least two lines of evidence that encourage exploration of the social dimensions of sustainability. The first is revealed in the interesting relationship between income (consumption) and well-being.

Available data show that life expectancy initially rises rapidly with per capita income but then levels off and is virtually flat between $10,000 and $25,000. It appears that 90 percent or more of the gain in life expectancy is "purchased" by the time income reaches $7,000–$8,000 per annum[9] (The World Bank, 1993). Similar relationships hold for related health and social indicators.

Seven to eight thousand dollars is only one-third to one-half of the per capita income of the world's wealthier countries. It seems clear, therefore, that quite substantial reductions in consumption by people in these countries might well be possible before there would be any significant deterioration in human welfare as measured by standard "objective" indicators. We should also note that various studies show that subjective "happiness" or well-being is not correlated with income in the upper income range. Indeed, people's perception of their own social and health status seems more a factor of relative social position than of absolute material wealth.

These data pose a serious challenge to conventional assumptions about the social need for continuous economic growth. They suggest that a healthy and sustainable society may, in fact, be possible at relatively modest income levels, even without any dramatic restructuring of society or social relationships.

The Case of Kerala

An example of the second argument for investing in social capital can be found in the state of Kerala, India. With an annual income per capita of only U.S.$350, the population of Kerala has achieved a life expectancy of 72 years (the norm for states earning $5,000 or more per capita), a fertility rate of less than two, and a high-school enrollment rate for females of 93%. According to Alexander (1994), "Extraordinary efficiencies in the use of the earth's resources characterize the high life quality behavior of the 29 million citizens of Kerala." Similarly, Ratcliffe (1978, p. 140) claims that Kerala refutes "the common thesis that high levels of social development cannot be achieved in the absence of high rates of economic growth. Indeed, the Kerala experience demonstrates that high levels of social development—evaluated in terms of such quality of life measures as mortality rates and levels of life expectancy, education and literacy, and political participation—are consequences of public policies and strategies based not on economic growth considerations but, instead, on equity considerations."

The point here is not to suggest that Kerala, with its unique political and cultural history, is a model for others to follow in detail. Rather, it is simply to emphasize that every society and culture is in part a social construction, not entirely a product of natural laws. In short, *there is nothing sanctified about our high throughput industrial culture*. Kerala shows that a high quality of life with minimal impact on the earth is possible through the accumulation of social rather than manufactured capital. As such, it is a sign of hope that other people in other cultures—possibly even in the global village as a whole—may also be able to organize in ways that distribute nature's limited bounty more equitably. There is no intrinsic reason we cannot learn to live sustainably in a low-throughput economic steady state.

Policy Implications

Empirical evidence suggests that the economy has already exceeded carrying capacity, yet we seem more determined than ever to address the problems of sustainability and persistent poverty through a new round of material growth. This often seems like a convenient way to avoid addressing inequity through policies to redistribute wealth, and it is a potentially dangerous path. It depends on the assumption that technological efficiency gains alone will succeed in reducing the human ecological footprint on the earth, even as the consumption of goods and services rises as much as tenfold.

At the same time there is clear evidence that meaningful social relationships and supportive community infrastructure may be more effective than technology in reducing the demand for energy and material. Policy makers would therefore be well advised to consider the "soft" alternative toward sustainability. Some relevant questions include:

- How can the state facilitate the shift in personal and social values implicit in a more caring society?
- What circumstances facilitate the development of sharing and mutual aid as a mode of life, even in the face of material scarcity?
- What kinds of formal and informal social relationships enhance people's sense of self-worth and personal security?
- Which of these personal relationships and community qualities reduce the compulsion to consume and accumulate private capital? In other words, what forms of social capital can substitute for manufactured capital?
- What sorts of policies would facilitate the development of these forms of social capital?

So far, material industrial society has avoided such questions in the production–consumption debate. However, addressing these issues would contribute not only to ecological sustainability but also to filling the spiritual void and general social malaise that increasingly seem to plague high-income, high-consumption societies.

Addressing the Double Bind of Sustainability

Humankind now seems to be the victim of a global catch-22 of its own making. More material growth, at least in the poor countries, seems essential for socioeconomic sustainability, yet any global increase in material throughput is ecologically unsustainable. What does ecological footprint analysis have to say about this double bind and how might we get out of it? One can draw several conclusions from the above analysis that address one or both sides of the dilemma:

- The wealthy already consume on average three times their fair share of sustainable global output. Since additional material growth in rich countries would appropriate additional carrying capacity, further reducing the ecological space available to poor countries, it is both ecologically dangerous and morally questionable. To the extent we can create room for growth, it should be allocated to the third world.
- Confidence in the ability of unregulated trade and technology to overcome ecological limits on material growth cannot be justified. Indeed, it is arguable that under prevailing assumptions, expanding trade and dominant technologies are allowing humanity to dangerously overshoot long-term global carrying capacity.
- Trade has been a major contributor to increasing gross world product in recent years. However, trade is one of the mechanisms by which the rich appropriate carrying capacity and increase their own ecological footprints, and, to the extent that trade increases total human load on the ecosphere and accelerates the depletion of natural capital, it reduces the ecological safety net for all and brings us closer to global limits. Global terms of trade must therefore be reexamined to ensure that it is equitable, socially constructive, and confined to true ecological surpluses. At the very least, prices must reflect ecological externalities, and the benefits of growth from trade should flow to those who need them most (see Rees, 1994a).
- On a finite planet, ecological trade is a zero-sum game—there can be no net importation of carrying capacity for the world as a whole. Ecological footprint analysis provides a useful tool for the development of regional ecological (i.e., physical) accounts. These would assist countries or (bio) regions to compute their true ecological loads on the ecosphere and to monitor their ecological/thermodynamic

trade balances. Such accounts would also enable the world community to ensure that aggregate global flows do not exceed sustainable natural income (global carrying capacity).

- Urbanization, globalization, and trade all reduce the negative feedback on local populations from unsustainable land and resource management practices. (For example, trade enables us to discount the value of local natural capital and blinds us to the negative consequences of our overconsumption, which often accrue in distant export regions.) This provides a further argument to shift the emphasis in development from global economic integration and interregional dependency toward intraregional ecological balance and relative self-reliance. (If all regions were in ecological steady state, the aggregate effect would be global stability.) This position is compatible with Daly and Goodland's (1993) recommended alternative "default position" on international trade: We should strive "to reduce rather than increase the entanglement between nations."

- Ecological footprint analysis supports the argument that to be sustainable, economic growth must be much less material and energy intensive than at present (see, for example, Pearce ,1994). It therefore supports the case for ecological tax reform in aid of resource conservation (von Weizsäcker, 1994). For example, depletion taxes and marketable quotas on natural capital inputs to the economy would (1) stimulate the search for more materially and energy efficient technologies; (2) preempt any resultant cost savings, thereby preventing the economic benefits of efficiency gains from being redirected to additional or alternative forms of consumption; and (3) generate an investment fund that could be used to rehabilitate important forms of self-producing natural capital (Rees, 1994b).

- Ecological footprint analysis provides a measure of both individual countries' ecological deficits and the global sustainability gap (box 3). The latter in particular is a measure of the extent to which the human economy must be dematerialized in order to fit within global carrying capacity. The present and related analyses confirm that a factor-10 reduction in the material and energy intensity per unit of economic service, as suggested by researchers at the Wuppertal Institute in Germany (Schmidt-Bleek, 1992a, b), is a reasonable if daunting goal.[10]

Conclusion

Appropriated carrying capacity and ecological footprint analysis provide several informative area-based indicators of sustainability. Unfortunately, these same indicators reveal that we are presently falling distressingly short of achieving that elusive goal. Such findings do not, however, support a counsel of despair. Rather, ecological footprint analysis raises a cautionary signal, suggests a variety of concrete sustainability guidelines, and supports a broad-based program of reforms that could redirect us in the direction we all seem to want to go. In short, to the extent that the assumptions and prescriptions of this approach are a better reflection of material reality that those of mainstream models are, the present analysis is a good-news story. The bad news is that most of the world seems committed as never before to the well-worn expansionist path.

References

BCSD (1993). *Getting Eco-Efficient*, report of the First Antwerp Eco-Efficiency Workshop, November 1993. Geneva: Business Council for Sustainable Development.

Catton, W. (August 18, 1986). "Carrying Capacity and the Limits to Freedom," paper prepared for Social Ecology Session 1, XI World Congress of Sociology, New Delhi, India.

Christensen, P. (1991). "Driving Forces, Increasing Returns, and Ecological Sustainability. In R. Costanza, ed., *Ecological Economics: The Science and Management of Sustainability*, pp. 75–7. New York: Columbia University Press.

Costanza, R., and Daly, H. (1992). "Natural Capital and Sustainable Development." *Conservation Biology* 1: 37–45.

Daly, H. (1986). "Comments on 'Population Growth and Economic Development.'" *Population and Development Review* 12: 583–85.

———. (1992). "Steady-State Economics: Concepts, Questions, Policies." *Gaia* 6: 333–38.

Daly, H., and Goodland, R. (1993). *An Ecological-Economic Assessment of Deregulation of International Commerce under GATT*. Discussion draft. Washington, D.C.: World Bank.

Elkins, P., and Jacobs, M. (1994). "Are Environmental Sustainability and Economic Growth Compatible?" in *Energy—Environment—Economy Modelling,* discussion paper no. 7. Cambridge, England: Department of Applied Economics, University of Cambridge.

Folke, C., Larsson, J., and Sweitzer, J. (1994). Renewable Resource Appropriation by Cities," paper presented at "Down To Earth: Practical Applications of Ecological Economics," Third International Meeting of the International Society for Ecological Economics, San José, Costa Rica, October 24–28, 1994).

Hannon, B. (1975). "Energy Conservation and the Consumer." *Science* 189: 95–102.

Hardin, G. (1991). "Paramount Positions in Ecological Economics." In R.

Costanza, ed., *Ecological Economics: The Science and Management of Sustainability*, pp. 47–57. New York: Columbia University Press.

Jaccard, M. (1991). "Does the Rebound Effect Offset the Electricity Savings of Powersmart?" discussion paper for BC Hydro. Vancouver: BC Hydro.

Kirchner, J., Leduc, G., Goodland, R., and Drake, J. (1985). "Carrying Capacity, Population Growth, and Sustainable Development." In D. Mahar, ed., *Rapid Population Growth and Human Carrying Capacity: Two Perspectives*, staff working paper no. 690, Population and Development Series. Washington, D.C.: World Bank.

Meadows, D. H., Meadows, D. L., and Randers, J. (1992). *Beyond the Limits.* Toronto: McClelland and Stewart.

Overby, R. (1985). "The Urban Economic Environmental Challenge: Improvement of Human Welfare by Building and Managing Urban Ecosystems," paper presented in Hong Kong to the POLMET '85 Urban Environmental Conference. Washington, D.C.: World Bank.

Pauly, D., and Christensen, V. (1995). "Primary Production Required to Sustain Global Fisheries." *Nature* 374: 255–57.

Pearce, D. (1994). "Sustainable Consumption through Economic Instruments." Paper prepared for the Government of Norway Symposium on Sustainable Consumption, Oslo, January 19–20, 1994.

Rees, W. (1988). "A Role for Environmental Assessment in Achieving Sustainable Development." *Environmental Impact Assessment Review* 8: 273–91.

———. (1990). "Sustainable Development and the Biosphere." Teilhard Studies no. 23. American Teilhard Association for the Study of Man, "The Ecology of Sustainable Development." *The Ecologist* 20, no. 1: 18–23.

———. (1992). "Ecological Footprints and Appropriated Carrying Capacity: What Urban Economics Leaves Out." *Environment and Urbanization* 4, no. 2: 121–30.

———. (1994a). "Pressing Global Limits: Trade as the Appropriation of Carrying Capacity." In T. Schrecker and J. Dalgleish, eds., *Growth, Trade, and Environmental Values*, pp. 29–56. London, Ontario: Westminster Institute for Ethics and Human Values.

———. (1994b). "Sustainability, Growth, and Employment. Toward an Ecologically Stable, Economically Secure, and Socially Satisfying Future," paper prepared for the IISD Employment and Sustainable Development Project. Winnipeg, Manitoba: International Institute for Sustainable Development. (Revised version in *Alternatives* 21, no. 4 [October–November 1995)].

———. (1995). "Achieving Sustainability: Reform or Transformation?" *Journal of Planning Literature* 9: 343–61.

Rees, W., and Wackernagel, M. (1994). "Ecological Footprints and Appropriated Carrying Capacity: Measuring the Natural Capital Requirements of the Human Economy." In A.-M. Jansson, M. Hammer, C. Folke, and R. Costanza, eds., *Investing in Natural Capital: TheEcological Economics Approach to Sustainability*, pp. 362–90. Washington, D.C.: Island Press.

RIVM (1991). *National Environmental Outlook, 1990–2010.* Bilthoven: Rijksinstituut voor Volksgezondheid en Milieuhygiene.

Saunders, H. D. (1992). "The Khazzoom-Brookes Postulate and Neoclassical Growth." *Energy Journal* 13, no 4: 131–48.

Schmidt-Bleek, F. (1992a). "MIPS—A Universal Ecological Measure." *Fresenius Environmental Bulletin* 1: 306–11.

———. (1992b). "MIPS Revisited." *Fresenius Environmental Bulletin* 2: 407–12.

Schneider, E., and Kay, J. (1994). "Life as a Manifestation of the Second Law of Thermodynamics." Preprint from *Advances in Mathematics and Computers in Medicine.* Waterloo, Ontario: University of Waterloo Faculty of Environmental Studies, Working Paper Series.

Sterrer, W. (1993). "Human Economics: A Non-Human Perspective." *Ecological Economics* 7, 183–202.

Vatn, A., and Bromley, D. W. (1993). "Choices without Prices without Apologies." *Journal of Environmental Economics and Management* 26: 129–48.

Vitousek, P., Ehrlich, P., Ehrlich, A., and Matson, P. (1986). "Human Appropriation of the Products of Photosynthesis." *BioScience* 36: 368–74.

von Weizsäcker, Ernst U. (1994). *Earth Politics.* London: Zed Books (see chapter 11, "Ecological Tax Reform").

Wackernagel, M. (1994). *The Ecological Footprint and Appropriated Carrying Capacity: A Tool for Planning toward Sustainability.* Unpublished Ph.D. thesis, University of British Columbia School of Community and Regional Planning, Vancouver.

Wackernagel, M., and Rees, W. (1996). *Our Ecological Footprint: Reducing Human Impact on the Earth.* Stony Creek, Conn: New Society Publishers.

WCED (1987). *Our Common Future.* World Commission on Environment and Development. Oxford: Oxford University Press.

Young, J., and Sachs, A. (1994). *The Next Efficiency Revolution: Creating a Sustainable Materials Economy,* Worldwatch paper no. 121. Washington, D.C.: Worldwatch Institute.

Notes

1. According to orthodox theory, free trade is invariably good, resulting in improved living standards and increased aggregate productivity and efficiency—increased carrying capacity—through comparative advantage.
2. Global fishery yields have fallen since 1989.
3. This input to the economy from nature is the "natural income" referred to in box 1.
4. Rebound effects can be avoided if adequate stock depletion taxes or marketable resource quotas are imposed. (Such incentives should be used to stimulate conservation in the first place.) Ecological taxation would raise unit resource prices, effectively capturing any efficiency savings and preventing their further circulation in the economy. However, because of

reduced material and energy intensity, consumer prices for goods and services would increase less rapidly than resource prices (Rees, 1994b).

5. An additional 0.74 ha of continental shelf "seascape" is appropriated to produce the average Canadian's annual consumption of 24 kg of fish.

6. The Vancouver Regional District (metropolitian area), with 1.6 million inhabitants and a land base of 2,930 km², has an ecological footprint of 6,720,000 ha, twenty-three times its geographic area.

7. Most of the imported "food" is fodder for domestic livestock. This is a sufficient "second law" explanation of the fact that animal manure represents one of the most pressing waste disposal problems confronting the Netherlands.

8. We should remember Liebig's law of the minimum in this context. The productivity and ultimately the survival of any complex system dependent on numerous essential inputs or sinks are limited by that single variable in least supply.

9. Figures in international dollars based on purchasing power.

10. "Reasonable" because a reduction in throughput of this magnitude seems necessary; "daunting" because a reduction of this magnitude through material efficiency alone seems impossible, at least within the next few decades. Sustainability may require that competitive individualism and the consumer lifestyle may give way to cooperative mutualism and an economy of sufficiency.

U.S. Consumption and the Environment

Michael Brower and Warren Leon*

Americans have long been proud of their material lifestyle—the two-car family, steaks on the grill, countless home gadgets and appliances. No society has ever reached such a high level of material consumption as our own, a fact that is taken by many Americans as proof of the superiority of our free-market system.

But American consumption habits impose a severe cost on the environment.[1] Our love of cars, big homes, and all the rest has placed the atmosphere, coastal waters, rivers, forests, and soils under siege. What's more, with much of the rest of the world aspiring to this lifestyle, the earth's life-support systems are in danger of being overwhelmed by a rising tide of consumption.

The growing environmental cost of consumption forces us to confront some difficult questions. How should we change our behavior so that citizens of poor countries can share in the fruits of modern industry—without wrecking the planet? How can we learn to use technology wisely so that the whole world may live comfortably within nature's means? The fate of the earth—and of our way of life—may well rest upon our answers to such questions.

---

*Michael Brower is president of Brower & Company, an environmental consulting firm in Andover, Massachusetts. Warren Leon is deputy director for programs at the Union of Concerned Scientists in Cambridge, Massachusetts.

Affluence and Consumption

Anyone who watches the evening news is sometimes confronted by discomforting images: the outstretched hand of a starving child, teeming refugee camps beset by disease and despair. The images disturb us not only because of the suffering they depict but also because they make us conscious of our own good fortune to live in one of the richest countries on earth. Not that Americans know nothing of poverty and economic insecurity. All too many in our population are poorly housed and underfed. Even some middle-class Americans fear that their economic well-being is threatened and their standard of living is declining.

But the economic gulf that exists between most Americans and the billions of people living in impoverished developing countries is wide. Compare the United States with Bangladesh, one of the world's poorest countries. Average per capita income in Bangladesh at the start of this decade was $210 a year, the literacy rate among adults was 35 percent, average food consumption was 2,000 calories a day, and life expectancy at birth was around fifty-one years. Average income in the United States was $22,000 a year, literacy was almost universal, average food consumption was 3,700 calories a day, and life expectancy was around seventy-six years.[2]

The gulf between rich and poor countries is marked not only by these grim demographic contrasts but also by the scale and type of material consumption. Most North Americans—like most citizens of western Europe, Japan, Australia, and a few other countries—belong to the "consumer class," a group consisting of about a billion of the world's most affluent.[3] Compared to people living in poorer countries, the members of this group consume enormous quantities of all sorts of things: energy, metals, minerals, forest products, fresh water, and fish, grains, and meat. According to the Worldwatch Institute, a typical citizen of an industrial country uses three times as much fresh water, ten times as much energy, and nineteen times as much aluminum as a typical citizen of a developing country.

Even within the consumer class, Americans rank near the top in most categories of resource use. The average American uses twice as much fossil fuel as the average resident of Great Britain and two and a half times as much as the average Japanese. The United States alone produces and consumes one-third of the world's paper, despite having just 5 percent of the world's population and 6 percent of its forest cover. The sheer waste of materials in our society is astounding: the typical American discards nearly a ton of trash per person per year, two to three times as much as is disposed of by the typical western European.

Many of the things we casually throw out—old pots and pans, food leftovers, plastic and paper goods—would be regarded as valuable resources in poor countries.

Of course, not all Americans fall in the same consumer category. There are gradations of consumption within as well as across societies. Wealthier people tend to consume more—they travel farther and more often, ride in more gas-guzzling cars, live in larger houses, discard more waste, and so on. And even the poorest developing country has its own wealthy consumers, eager to spend their money on many of the same goods Americans do. Consumerism is not just a facet of American life but a burgeoning worldwide trend.

The Environmental Costs of Consumption

Everything people consume has an impact on the environment—taking even a single branch from a forest leaves some mark on the ecosystem. And yet, not all consumption is bad. For one thing, some items are essential to human existence and well-being, whereas others are mere luxuries. In fact, the problem for the very poor both in the United States and in many other countries is that people consume too little of the essentials of life—food, fuel, and shelter—leading to hunger and deprivation. Moreover, some kinds of consumption—such as the burning of fossil fuels—are inherently harmful, whereas others—such as the use of forest products and the growing of food crops—if done wisely, can be sustained indefinitely without causing deterioration of the underlying resource base.

It is the nature and scale of consumption that matters, and much of the consumption in the United States is extremely damaging to the environment. The coal we burn to generate electricity produces particulates, acidic compounds, mercury, and other toxic materials that foul the air, soil, and water. The gasoline we burn to run our cars creates compounds that combine chemically to form smog. To supply our demand for wood products, the timber industry thins and clear-cuts millions of acres of national and privately owned forest annually; as a result, although the total area covered by forests in the United States is gradually expanding, the health and diversity of many forest ecosystems are severely degraded. Even our taste for meat is a serious problem, since the amount of grain used to feed livestock multiplies the impacts of intensive methods of agriculture on air, soil, and water, and the millions of cattle and other livestock create wastes that flow, often untreated, into rivers and lakes.

This is only the start of the list of harmful impacts linked to consumption patterns in the United States and other industrial countries. Moreover, the impacts often extend well beyond our own borders. For example, global warming (caused largely by emissions of heat-trapping gases from industrial countries) could result in a major drop in crop yields in areas of Africa and Asia that are already stricken by hunger. In addition, some of the products we use are produced in developing countries in environmentally harmful ways. Such trade brings obvious economic benefits to the exporting countries (although the benefits often accrue to a small number of wealthy individuals and corporations) but also causes terrible ecological damage, raising difficult trade-offs for developed and developing countries alike.

Putting the United States on a path toward sustainability means changing both our consumption habits and the way products are manufactured, delivered, and disposed of. In other words, it involves changing not only technology but also personal, corporate, and government behavior. To better understand how America's consumer society needs to change to protect the environment, we recently undertook, on behalf of the Union of Scientists, the first-ever comprehensive study of the wide range of consumer activities. We wanted to identify which of the things that consumers buy and do cause the most environmental damage, as well as changes that people could make at the personal, community, and government levels to reduce that damage.[4] The most important conclusion we reached is that only a few consumer activities are responsible for the vast majority of consumer-related damage. We grouped spending into fifty major categories (e.g., dairy products, furnishings, clothing) but found that most environmental degradation is linked to just seven of those categories. Collectively, the dirty seven are responsible for roughly two-thirds of the greenhouse gases that can be traced to consumer behavior and three-fifths of the air pollution, as well as most of the changes to natural habitats and considerable water pollution.

This finding is important because it shows that how people spend their money does indeed matter. Nothing would be more discouraging than to discover that no matter what a person does (short of living in a cave), about the same amount of environmental damage will result. Instead, some kinds of consumption are *much* worse for the environment than others.

The Car Problem

Our analysis shows that the use of cars and light trucks (such as minivans and sport utility vehicles) causes more environmental damage than anything else consumers do. This is not surprising since nothing symbolizes the excesses of America's consumer society better than our long-standing love affair with the automobile. Starting early in this century, cars have become entrenched in our way of life. Suburbs are built around them; shopping malls and "drive-thru" restaurants cater to them. Thanks in part to the power of advertising and movies, they have come to represent many of the things Americans aspire to most, including individual freedom, mobility, sexual attractiveness, and social status.

Americans own more cars and drive them longer distances than just about anyone else.[5] The rate of car ownership in the United States—one for every two people—is 39 percent higher than that in Britain and 78 percent higher than that in Japan. On the basis of vehicle-miles traveled per person, Americans drive twice as much as the British and over three times as much as the Japanese. Comparisons to developing countries are even more striking. Our rate of car ownership is 7 times greater than that of Mexico, 175 times greater than that of India, and more than 300 times greater than that of China.

It is reasonable to ascribe at least some of the differences between U.S. transportation habits and those of other countries to geography: the United States, after all, is a very large country with a relatively low average population density, so one can expect that people would drive more. Although geography certainly plays a role, it is not the only factor. The former West Germany, for example, has about the same population density as Massachusetts, but West Germans drive about 20 percent less than do residents of Massachusetts. And in Canada, with an even lower population density than the United States, public transportation is used five times as often in urban travel. These differences reflect differing national policies rooted in history and culture. Western European countries and Japan, for example, have put much higher sales taxes on both gasoline and automobiles than has the United States and have provided more consistent and generous support for public transportation and other alternatives such as bicycling and walking.

All of the driving Americans do has a very damaging impact on the environment. Cars account for just over a quarter of the fourteen tons of carbon emissions (the leading cause of global warming) generated by the average household (both directly and in the manufacturing of the products households buy). While a small share of this impact can be traced to automobile manufacturing, driving accounts for most (85 percent).

Similarly, much of the air pollution consumers cause comes from cars and light trucks: 22 percent of common air pollutants (ozone, sulfur dioxide, fine particulate matter, and the like) and 46 percent of toxic air pollutants such as benzene and formaldehyde. Indirectly, vehicles are also a major source of water pollution, from automobile manufacturing, runoff from highways, and oil and gasoline production. They are even responsible for 13 percent of the ecologically harmful land use—for the road network.

Because cars cause so much harm, quite modest changes can produce a significant reduction in people's overall environmental impact. For example, if their cars averaged 27 rather than 23 miles per gallon, the average household's total contribution to global warming and air pollution would decrease by 5 percent. Obviously, regularly carpooling, taking public transportation, bicycling to work, or making other more dramatic changes would produce an even greater reduction.

The Rest of the Top Seven

In addition to automobiles, six other consumer spending categories cause especially heavy environmental damage. Here are short summaries of how they impact the environment.

Meat and poultry. Meat and poultry production requires large amounts of water and causes 20 percent of the common (as opposed to highly toxic) water pollution related to consumer expenditures. It also uses a significant share of the nation's land—800 million acres for grazing livestock and an additional 60 million acres to grow animal feed. Red meat causes especially high amounts of environmental damage for the nutrition it delivers.

Fruit, vegetables, and grains. Irrigated crops grown to meet consumer demand use an enormous quantity of water (30 percent of consumer-related water use). The pesticides and fertilizers used on food crops cause 5 percent of consumer-related toxic water pollution. Food crops also use substantial amounts of land.

Household appliances and lighting. Electricity seems clean and nonpolluting when it's used in the home, but most of it is generated by burning polluting fossil fuels, especially coal. Appliances and lighting are responsible for 15 percent of the greenhouse-gas emissions and 13 percent of the common air pollution related to consumer expenditures.

Home heating, hot water, and air conditioning. Cooling and heating homes and water have an impact on global warming and air pollution similar to that of appliances and lighting. Systems that rely on electricity or oil contribute heavily to both problems. Most fireplaces and wood stoves are especially high air polluters.

Home construction. The land and wood used for homes are responsible for about a quarter of consumers' impact on wildlife and natural ecosystems. Six percent of consumer-related water pollution comes from manufacturing the materials for new homes and disturbing the soil during construction.

Household water and sewage. Despite advances in sewage treatment, municipal sewage remains a major source (around 11 percent) of water pollution, especially affecting coastal areas and estuaries. Interestingly, households' home water use is only 5 percent of the total, compared with nearly 74 percent for food production and distribution.

The Role of Technology

Knowing what causes the most damage is, of course, essential, but it doesn't necessarily lead to environmental improvement. After all, many of the things that cause the most damage are pretty fundamental to a middle-class lifestyle that most Americans would like to retain. To improve the environment while preserving their familiar way of life, many people pin their hopes on better technology. And technology can indeed go a long way toward mitigating the impacts of consumption.

Encouraging trends are already under way. Today's new cars have twice the fuel efficiency as models available in 1973, for example, and are considerably safer, to boot. Similarly, over the past generation, refrigerators, washing machines, and other appliances have been designed to use much less electricity while providing better performance. Most industrial processes have also become less energy intensive and wasteful. And a combination of emissions controls and fuel switching has substantially reduced emissions of several forms of pollutants, with the result that many cities have seen constant or improving air quality in the past two decades.

Clearly, much more can be done in this direction. Cars have been designed and tested that travel seventy to eighty miles on a gallon of gasoline—more than three times the U.S. national average today. Fuel-cell and electric vehicles offer the potential to eliminate pollution in transportation, if the electricity is generated from wind, solar, or hydroelectric power systems.[6]

Crops can be grown in an ecologically sound fashion while achieving long-term yields and costs comparable or superior to those seen in conventional agriculture. In one example, the Spray farm in Knox County, Ohio, there has been no application of artificial fertilizer or herbicides in over fifteen years, yet average yields of corn, soybeans, wheat, and oats in the mid-1980s were higher than the averages for other farms in the same county. On a much larger scale, Gallo and some

other wine companies have been so persuaded of the benefits of ecologically sound farming that they have converted their vineyards to this
approach. Sustainable agricultural practices include crop rotations to
minimize soil nutrient loss and reduce pests, contour plowing to reduce
erosion, and the integration of crops with livestock production to take
advantage of natural fertilizers.[7]

But improving technology can take us only so far. One problem is
that technological change is often at least partially offset by a general
rise in both population and per capita consumption. For example, the
total number of miles driven by Americans increased about as fast as
automobile fuel efficiency in the 1980s, with the result that total gasoline consumption stayed about the same. Another problem is that
promising technological "solutions" can sometimes create new and
unexpected problems. As an example, chlorofluorocarbons were once
seen as the preferred alternative to ammonia in refrigerators and air
conditioners—until their role in the depletion of the stratospheric ozone
layer was discovered.

Lastly, there is the rest of the world to consider. We Americans—275
million of us—share the earth with 5 3/4 billion other people, the vast
majority of whom do not enjoy the material luxuries we do. Consider
the potential for growth in energy use alone if the entire world emulates
our present consumption habits: If world population stabilized at
around 10 billion—a relatively optimistic forecast—total consumption
of energy could be ten times what it is today. As this suggests, Americans must set a better example if the world is ever to live within the limits of the environment and natural resources.

Toward Sustainability

The long-term solution to environmental degradation lies in the concept of sustainability. Although difficult to define precisely, the basic
idea is to use the earth's resources to meet our needs in such a way that
future generations will be able meet their own needs at a comfortable
standard of living. For the United States, it means, above all, reducing
consumption of those items that cause the most environmental damage
and choosing consumer goods that will have minimal impact on the
environment. Sustainability also involves ensuring equity in the distribution of environmental impacts and risks among all Americans. Just as
the United States must not demand an unfair share of the world's
resources or generate more than its share of the world's pollution,
wealthier Americans must not rob poorer Americans of their rightful
share of clean air, clean water, and other environmental necessities. At

present, poor communities—especially communities of color—are far more likely than other communities to be sites of major polluting industries, hazardous waste dumps, and waste-treatment facilities. Thus, middle- and upper-income people often do not see the consequences of their own consumption habits.

Various strategies need to be followed to make sustainability a reality. Some are at the personal level: As citizens of a global community, we need to cultivate a personal ethic of material modesty and environmental consciousness. In practical terms, this means making a commitment to consume fewer of those things that damage the environment the most.

In general, Americans should pay particular attention to four types of consumption—energy, transportation, food, and hazardous chemicals. Among other things, this means buying cars with the highest available fuel efficiency and finding ways to drive less; saving energy by insulating and weatherproofing our homes; purchasing high-efficiency appliances, lamps, and heating systems; cutting back on consumption of meat, and buying certified organic foods; reducing the use of toxic household products such as lawn pesticides and paint remover; and generally avoiding buying material-intensive goods we do not actually need—that second car, third television, or overly large house.

Some of these actions could entail changes in personal behavior and lifestyle, but none has to mean sacrificing personal wealth or happiness. In fact, many of the things we can do as individuals to help protect the environment—eating less meat, walking or bicycling more—can directly enhance our health and sense of well-being. And reducing expenditures on some material goods may free up money and time for other, perhaps more rewarding activities, such as learning a new hobby, participating in a community organization, or spending time at home with one's family.

But we cannot rely solely on voluntary action by individual consumers to solve environmental problems, since consumers do not have complete control over what they consume and how much damage it causes. For example, those people thirty years ago who were concerned about the health hazards from lead in gasoline had to wait for oil companies to start producing unleaded gas.

Even when individual consumers theoretically have an environmentally sound choice, they may not view it as a practical alternative. An individual who drives twenty miles to work is unlikely to switch to mass transit if the only available bus takes twice as long, comes only once an hour, and has worn, ripped, graffiti-covered seats. Homeowners are unlikely to purchase benign nonchemical forms of lawn pest control if none of the stores in their town display such products and the sales clerks do not even know how they can be ordered.

In many cases, therefore, what needs to change is the choices available to consumers. The key decisions then need to be made at the corporate, institutional, or government level rather than among individuals. Americans seeking to reduce the environmental impact of products would often be best served by pressuring their local, state, or national government to adopt policies that make it easy, or even a requirement, for manufacturers and users of products to choose the environmentally sound option. For example, the government decision to require appliance manufacturers to list the energy costs of their products not only provided consumers with useful information but gave manufacturers a reason to improve the energy efficiency of their products. Cities that set up curbside recycling programs increase citizen participation in recycling. In the case of leaded gasoline, it was obviously much more effective for the government to ban this dangerous product altogether than to wait for every manufactuer and every consumer in the country to voluntarily switch to the unleaded alternative.

We also have to remember that, paradoxically, consumers are not the only ones who consume things, even though discussions of consumption–environment connections most often focus on "consumer goods" and what individuals consume. Businesses, organizations, and governments also consume things as part of their activities, and sometimes they are more responsible for pollution than are individuals. We therefore should not assume that the decisions of individual consumers cause most environmental damage. Instead, we should focus some of our attention on changing organizations.

Nevertheless, the role of individual Americans and their personal consumption choices remains large. Just as we vote with our ballots on election day, we vote with our dollars when we choose to buy or not to buy particular products. Not only do we send important messages to manufacturers when we buy their products, but we let our family, friends, and neighbors know something about our values.

Notes

1. We discuss the relationship between American consumption and the environment more fully in Michael Brower and Warren Leon, *The Consumer's Guide to Effective Environmental Choices: Practical Advice from the Union of Concerned Scientists* (New York: Three Rivers Press, 1999). Two other useful but very different book-length introductions to the relationship between consumption and the environment are Alan Durning's popular overview, *How Much Is Enough?* (New York: W.W. Norton, 1992)

and a collection of brief summaries of scholarly books and articles by Neva R. Goodwin and others, *The Consumer Society* (Washington, D.C.: Island Press, 1997).

2. For comparative statistics on human and material conditions in various countries, see World Resources Institute, *World Resources 1998–99* (New York: Oxford University Press, 1998); the World Bank's annual *World Development Report*; and the annual reports of the United Nations Population Fund and the World Health Organization.

3. Alan Durning in *How Much Is Enough?* explains and discusses the concept of a global consumer class.

4. Brower and Leon, *Consumer's Guide to Effective Environmental Choices.*

5. For an overview of transportation in the United States, with comparisons to other countries, see Deborah Gordon, *Steering a New Course: Transportation, Energy, and the Environment* (Washington, D.C.: Island Press, 1991).

6. Jason Mark, *Zeroing Out Pollution: The Promise of Fuel Cell Vehicles* (Cambridge, Mass.: Union of Concerned Scientists, 1996).

7. For an introduction to sustainable agriculture, see Margaret Mellon et al., *Sustainable Agriculture: A UCS Briefing Paper* (Cambridge, Mass.: Union of Concerned Scientists, 1995).

Part III
RELIGION AND THEOLOGY

Overview of Religious Perspectives

Audrey R. Chapman

Increasing awareness of the environmental crisis has prompted a reevaluation of the potential role of the religious community in dealing with environmental problems. The essays in part 1 of this volume provide a brief history of the science and religious dialogue on environmental issues and outline the potential contributions of the religious community. This section considers the perspective of the traditions—Christian, Jewish, and Muslim—that were represented at the November 1995 conference "Consumption, Population and the Environment: Religion and Science Envision Equity for an Altered Creation." It also includes the text of a statement of an international and interfaith consultation convened in conjunction with the 1994 United Nations International Conference on Population and Development and a topical treatment of the religious ethic of frugality. A review of these sources underscores both the strengths and weaknesses that many faith communities have in addressing population and consumption issues.

In considering this topic, it should be noted from the outset that most faith communions have only recently begun to deal with threats to the environment from overpopulation and overconsumption and to examine what their traditions have to say about these issues. Moreover, when the religious community addresses these and related issues, it does not speak with a single voice. Given the complexity of the issues and the significant variation in traditions, differences are to be expected. It is a hopeful sign that at least the leadership and some religious thinkers are beginning to address population and consumption problems in a serious way. Nevertheless, members of the religious community have yet to invest the time and attention accorded the environmental crisis (in the

111

abstract) or economic justice issues. Nor have most faith groups been willing to confront the problematic features of their own traditions.

Ecological Awareness

Some thirty years ago, Lynn White published a seminal article that subsequently played an important role in engendering ecological awareness in the religious community. In the article, White, a medieval historian, implicates Christianity as a major contributor to the environmental crisis. According to White, the biblical mandate to dominate nature, along with an anthropocentric Christian orientation, resulted in an instrumental rather than a respectful approach to nature and furnished an impetus to the development of an environmentally destructive science and technology.[1] This has been a much debated subject ever since; and many analysts have disagreed with White's thesis or at least have qualified it. Nevertheless, White's criticism of Christianity played a constructive role in prompting a reevaluation and reformulation of Christian theology and ethics to be more sensitive to environmental concerns.

During the past thirty years, the environmental crisis has stimulated a "greening" of religious thought as thinkers across the religious spectrum have begun to respond in a meaningful way to a growing sense of the fragility, vulnerability, and interdependence of the creation. Benedicta Musembi and David Anderson's survey "Religious Communities and Population Concerns" concludes, for example: "Religious communities, especially at the leadership levels, are showing an increasing concern for issues of consumption, population, and environmental sustainability."[2] Another recent review argues that the challenge to integrate ecology and justice concerns with the Christian faith has transformed Christian self-understanding in a relatively short period.[3] Theologians and ethicists from a wide range of faiths are reexamining their scriptural traditions and symbolic heritage in an effort to highlight human responsibility for earth stewardship and ecological justice. Whereas thirty years ago White considered Christianity to be a cause of the problem, many scientists, secular environmental thinkers, and environmental activists now look to the religious community with a sense of common purpose and commitment.

Nevertheless, while many religious traditions have resources, few have a fully developed systematic environmental ethic relevant to contemporary issues. Moreover, religious communities differ quite considerably in their commitments to respond to the environmental challenge. Daniel Fink's comments on Judaism can apply more broadly across the religious spectrum:

> For most of Jewish history, our sacred texts—from Hebrew Scriptures to Talmud to medieval philosophical, legal and mystical literature—have dealt with ecological issues incidentally, as they arose. Ecology was not a discrete area of inquiry; it was, instead, an integral part of the weave of relationships between God, humanity in general (and Israel in particular) and the rest of the nature world. Furthermore, Jewish positions on environmental issues have never been monolithic. In this, they reflect the multi-vocal nature of our tradition's texts and worldviews.[4]

Various essays in this volume reflect efforts to explore the contributions of particular religious traditions to preserving and nurturing God's good creation. Michal Smart discusses three elements within Jewish tradition: God's perceived ownership of the world; the ethic of *bal taschit*, which literally means "do not destroy"; and *tzedakah*, the Hebrew word for justice. Likewise, in their essay, Abdul Cader Asmal and Mohammed Asmal emphasize the Islamic belief that everything emanates from God and that humankind serves as God's trustee or vice regent on earth with the responsibility to respect and preserve God's creation according to God's laws. Emmanuel Clapsis describes the heightening Orthodox interest as manifested by the "Encyclical on the Environment" issued by the ecumenical patriarchate in 1989 to raise the awareness of all Orthodox Christians concerning the ecological crisis. His essay contains the text of proposals made by participants in an interreligious and interdisciplinary symposium, "Revelation and the Environment: A.D. 95–1995," organized by Ecumenical Patriarch Bartholomew to mark the nineteen-hundredth anniversary of the composition of the final book of the Christian Bible. A notable outcome of that conference was a redefinition of the concept of sin to include activities that pollute the environment or abuse nature.

One of the formative challenges for the religious community is to develop effective instruments of religious education to instruct, inspire, and transform the moral life of members to live in accordance with a sustainable ethic. The Environmental Justice Program of the (U.S.) National Conference of Bishops, described in David Byers's essay, is the most comprehensive of several national denominational efforts. Its goal is to "help Catholics in parishes, diocesan institutions and organizations, and national organizations to take seriously the moral obligation to care for God's gift of creation." To that end, the program has a wide range of activities, including the development of resources and the provision of small grants to support parish and diocesan initiatives that can serve as models for others.

Recently a series of ten conferences on religion and ecology was held at the Center for the Study of World Religions at Harvard University, coordinated by John Grim and Mary Evelyn Tucker of Bucknell University. Each of the conferences explored the particular intellectual and symbolic resources of a specific religious tradition regarding views of nature, ritual practices, and ethical constructs in relation to nature. The conferences focused on virtually all of the world's major religious traditions—Judaism, Islam, Christianity, Confucianism, indigenous traditions, Buddhism, Hinduism, Jainism, Daoism, and Shintoism. Commissioned papers will provide the basis for a volume for each of the conferences, resulting in a veritable library. Perhaps more than any other development, these ten conferences attest both to the stirrings of ecological consciousness within the religious community and to how far away most of these communities are from formulating a systematic environmental ethic.

Population

The population crisis constitutes a serious challenge to many religious traditions. With the exception of Buddhism and some aboriginal traditions, the heritage of most major religions can be characterized as solidly "pro-natal."[5] Formulated in earlier periods when populations were small, environmental impacts relatively minimal, and survival difficult, their teachings often emphasize the obligation to have children. Many religions have commandments that define the perpetuation of one's family lineage as a religious duty. Some traditions also view reproduction as essential to being an adequate human being.[6] In one of the best-known texts, the first chapter of the Hebrew Bible (Old Testament), the first injunction given to humankind after its creation is to procreate. "Be fruitful and multiply, and fill the earth and subdue it."[7]

Some religious communions have responded to changed circumstances by acknowledging a need for responsibility in making family-size decisions. As an example, the Clapsis essay mentions that most Orthodox theologians now encourage the use of effective birth control and smaller families. Other religious communities have resisted reformulating their traditions in the face of current population pressures and continue to promote large families, oppose the use of birth control, and reject the need for population policies. As the essays by David Byers and Abdul Cadar and Mohammed Asmal acknowledge, the official teachings of the Roman Catholic Church and interpretations of the Qur'an by many Muslim scholars have opposed the use of contraception and family planning. Like Catholicism and Islam, Hinduism has traditionally opposed fertility limitation, teaching that the purpose of marriage is the production of children.[8] Many members of the Southern Baptist Convention, the largest denomination in the United States, apparently consider overpopulation to be a myth.[9]

Historically, mainline Protestant denominations have been in the forefront of efforts to make the dissemination of birth control information and services legal, and several of these denominations have long had policy statements that support responsible family planning as a clear moral duty. More recently, several have adopted positions that include the goal of stabilizing the U.S. population. However, it is unusual for these denominations to link population and consumption issues or to acknowledge that American consumption patterns wreak at least as much havoc on the environment as high levels of population growth in other countries. Moreover, even denominations that endorse family planning and contraception can be uneasy with population policies that set quantitative limits. It should also be noted that the contentious abortion debate has tended to distract attention from and color the issue of global population pressure.

When evaluated through the perspective of the woman-centered approach to population adopted at the 1994 International (Cairo) Conference on Population and Development, which is described in several of the essays in this volume and in an excerpt from the *Programme of Action* represented in part 4, the patriarchal and sexist character of many religious traditions constitutes another significant barrier to lowering rates of population growth. A statement from "World Religions and the 1994 United Nations International Conference on Population and Development," which is the next piece in this section, describes this heritage as follows, "The witness of religions to the equal dignity, rights, and freedom of women and men has not been uniform or unambiguous." It goes on to note that "many ancient texts and modern interpretations reflect cultures of male dominance and are construed as divine authorization for the subjugation of women." The report observes that the advancement of women's rights and roles is currently being debated among and within religious communities and cultures, but many feminists would claim that this debate is too little and too late. Relevant to this issue, representatives of both the Roman Catholic Church and the Islamic community took positions at the Cairo Conference and the subsequent Fourth World Conference on Women held in Beijing in September 1995 that were widely considered to be unsympathetic, even hostile, to promoting women's rights and equality.

From the above, it is clear that there is no consensus within the religious community on this subject. It is therefore relevant to quote from the closing caution in the Musembi and Anderson's survey: "Religious com-

munities remain divided on principles, emphasis, and strategies, both
from each other and, at times, within their own constituencies. This is
particularly true of the complex personal and political issues involved in
the debate over the world's growing population and efforts to achieve
population stabilization."[10]

Consumption

In contrast with their problematic heritage on population issues, most
religions have an ethic that offers correctives to the exploitation and
destruction of the environment and the excessive consumption patterns of affluent people in developed and developing countries. Some
of the relevant points of convergence are principles of respect for
nature, injunctions not to waste, and commitments to social justice or
sharing with the less fortunate. Many have an incipient or explicit
ethic of interdependence with the rest of creation. The non-monothe-
istic religions tend to vest nature with greater intrinsic worth and to
stress more centrally the need for humanity to live in harmony with the
environment. Judaism, Christianity, and Islam traditionally image
humanity as the steward or custodian of nature. While historically
interpreted in a variety of ways, contemporary work on the concept of
stewardship usually emphasizes that humanity is the manager, responsible to God, the owner of the creation. Some expressions of the ethic
of stewardship teach that humans must not damage or abuse nature in
any way and that the use of natural resources must be balanced, not
excessive. Conservation is explicitly enjoined by the Qur'an, and there
are rules in Islamic law for the conservation of forests, water, and animals.[11]

According to "World Religions and the 1994 United Nations International Conference on Population and Development," that is included
in this volume, most religious traditions also have teachings that discourage overconsumption and criticize greed and lack of sharing. This
statement may have been the case historically. However, religions vary
in their contemporary application and emphasis. This can be seen in
their treatment of the value of wealth, which is intrinsically related to
the issue of overconsumption. For example, Christian scriptural teachings that denigrate materialism and emphasize spirituality and sharing
may have little to do with contemporary religious practice. While Jesus
emphasized that it is not possible to serve two masters, God and mam-
mon, many American Christians believe that wealth is a sign of God's
favor. One recent survey of the treatment of work, money, and materialism in popular Christian evangelical literature concludes that the

authors of these books tacitly endorse the structures and systems of secular American life. According to researcher Marsha Witten,

> The status quo of workplace hierarchies, with the logic of exchange that everywhere undergirds them, is not just accepted but sacralized. Incentives for spiritual behavior employ—and not just metaphorically—the language of the trading floor. The accumulation of money, except in extreme excess is deemed just fine if only one's attitude is right. And only the vaguest limits are placed on the acquisition and consumption of goods, although the appropriate contours of many other behaviors (e.g., how long a workweek one should have, or how to keep a record of one's prayer activities) are specified in detail.[12]

Various articles in both this and the next section reflect the commitment of the religious community to promoting a more sustainable future. Sustainability is an evolving term within both the scientific and the religious communities. Within the religious community, its meaning often reflects the theological assumptions, as well as the goals and agendas of those using the term. Several religious thinkers apply environmental stewardship mandates to more contemporary sustainability issues, but many religious communities fail to stress this linkage.

To date, the religious community has been timid in dealing with the implications of lifestyle and consumption issues for the environment. Historically, frugality—moderation, temperance, thrift, efficient usage—was a prime Christian economic norm. As put forward by James Nash in his essay on this subject, the concept of frugality provides a norm of economic activity for both individuals and societies that leads to ethically disciplined production and consumption for the sake of some higher ends. As such it is an instrumental value, not an end in itself. In our affluent and "sumptuous society," frugality is countercultural because it requires a sufficient reduction of the production and consumption of goods in order to make available the material conditions for essential economic and social development in poor nations. It is also potentially liberating in that it rejects the popular assumption that humans are ceaselessly acquisitive and resists the temptations of consumer promotionalism. Frugality also offers a vision of an abundant life that is earth affirming and enriching. Nevertheless, as Nash acknowledges, becoming frugal on a personal or a societal level will be difficult. Moreover, while frugality was once near the center of Christian economic ethics, it "has become one of the most neglected—and unnerving—norms in modern morality." That Nash's essay is the only one in this section that strongly advocates frugality exemplifies the reticence within the religious community to confront the issues of excess consumerism that are so central to the future of this planet. This has considerably weakened the witness and credibility of the religious community.

Michal Smart comments in her essay that in the final analysis intellectual exploration of religious tradition will not be sufficient to meet the challenges that confront us. She quite rightly concludes that if we, as a society, are to wean ourselves from our addiction to material consumption, we will need to develop our inner resources and achieve a sense of spiritual fulfillment and well-being. As she intuits, the environmental crisis is also a fundamental religious crisis. According to Smart, "We need a process of religion coming into being—coming to claim and utilize its full power to instruct and inspire, to engage in thoughtful and meaningful ways with issues of contemporary concern, and to inform the moral life of its community." Whether the religious community accepts this challenge may determine the well-being of the planet and the relevance of religion to the human future.

Notes

1. Lynn White, Jr., "The Historical Roots of Our Ecological Crisis," *Science* 155 (1967): 1203–7.
2. Benedicta Musembi and David E. Anderson, "Religious Communities and Population Concerns," 1995. This literature search and bibliography were prepared by the Population Reference Bureau for the Pew Charitable Trusts' Global Stewardship Initiative and can be obtained from the Population Reference Bureau, 1875 Connecticut Avenue N.W., Suite 520, Washington, D.C. 20009-5728.
3. Peter W. Bakken, Joan Gibb Engel, and J. Ronald Engel, "Critical Survey," in their (edited) *Ecology, Justice, and Christian Faith: A Critical Guide to the Literature* (Westport, Conn.: Greenwood Press, 1995), p. 3.
4. Daniel B. Fink, "Judaism and Ecology: A Theology of Creation," *Earth Ethics* 10 (Fall 1998): 6.
5. Harold Coward, "Introduction," in Harold Coward, ed., *Population, Consumption, and the Environment: Religious and Secular Responses* (Albany: State University of New York, 1995), p. 14.
6. Rita M. Gross, "Buddhist Resources for Issues of Population, Consumption, and the Environment," in Coward, ed., pp. 155, 163.
7. Genesis 1:28, *The Holy Bible,* New Revised Standard Version.
8. Klaus K. Klostermaier, "Hinduism, Population, and the Environment," in Cowan, ed., pp. 137–40.
9. Musembi and Anderson, p. 24.
10. Musembi and Anderson, p. 37.
11. Coward, "Introduction," pp. 14–19.
12. Marsh G. Witten, "Where Your Treasure Is: Popular Evangelical Views of

Work, Money, and Materialism," in Robert Wuthnow, ed., *Rethinking Materialism: Perspectives on the Spiritual Dimensions of Economic Behavior* (Grand Rapids, Mich.: Wm. B. Eerdmans Publishing, 1995), 140–41.

World Religions and the 1994 United Nations International Conference on Population and Development

In anticipation of the United Nations International Conference on Population and Development (ICPD), in Cairo, Egypt, September 5–13, 1994, thirty thinkers from the world's major religious traditions held an international and interfaith consultation in Genval, Belgium, in May 1994, in the belief that the world's religions could make a fruitful contribution to reflection on the themes, rationales, and recommendations of the ICPD's proposed Programme of Action.

Only a few conferees were officials of their religious communities and thus bound to put forward an official position. Most were religious thinkers in the mainstream of their traditions. All were asked to describe and explain the official stance of as well as the diverse contexts, mores, and positions represented within their faiths. Some expressed personal dissent from an official position of their tradition; others noted the extent to which the stances and practices of followers differ from the authoritative teachings. Each knew, however, that the readers of this document would be ill served unless both the authorized position and the range of opinions and practices of their respective faith traditions were depicted. To reflect that diversity of perspective within and among the faiths, the term religious communities *is often used in this document.*

Preface

The religious communities of the world have a large investment in issues connected with the words *population* and *development*. All would explain that the root of that concern lies in a witness to what is sacred. This is usually understood as something that transcends, that goes beyond, the bounds of ordinary human and natural existence. Most people of faith focus their witness on God or gods and thus describe their relationship to what is sacred in explicitly theistic language, while millions of others refer to what is sacred and their connection to it in more generalized, nontheistic terms. All are concerned to relate this religious dimension to the created order.

With such profound, religiously charged human concerns like population and development at stake, many believers feel compelled to offer religiously informed perspectives on these issues, and some attempt to persuade others to accept their point of view. Sometimes their sense of responsibility and commitment to their religious vision is so strong that they may seek to impose their beliefs on others. At their best, however, people of faith find occasion to address social crises by thinking and acting together, fully aware that no one religion, segment of a religion, or cluster of religions is to be privileged or to have dominance in the family of religious communions or in public debate. Further, they look for opportunities to work with people of goodwill who profess no explicit faith and do not consider themselves to be religious but pursue shared goals on other grounds.

The gathering in Belgium was not intended to make a contribution to the theory of interreligious dialogue. Instead, the intent was practical: to articulate the interests and witnesses of religious communities on the themes of the ICPD and to seek points of convergence on these urgent issues. The measure of consensus reached does not represent any kind of agreement in detail within or among religious communities. It does, however, provide a positive counterpoint to media accounts that have emphasized a few notable cases of religious opposition to the ICPD agenda and particular items on that agenda.

Recognizing and respecting the great diversity in religious beliefs and practices, the participants nonetheless affirmed that the faiths share a regard for human persons both as individuals and in communities. This regard includes but also surpasses mere biological existence: it encompasses the health and well-being of persons, their quality of life, and their dignity, rights, and equality in the context of society and culture. Underlying these concerns is a belief that persons have inherent value and are not to be treated as objects, as instruments for other purposes.

The ICPD provides an apt occasion, as people across the globe think and speak and work together, to assess and renew religious commitments to this fuller understanding of the human person.

It should be noted that if there were massive opposition from all religions, the ICPD would probably fail in implementing any program, no matter how impressive its statistics, how scientific its arguments, or how rational its proposed solutions. Religions, in other words, can have a powerful influence in guiding human thought and behavior.

Most religious believers and communities who are informed about the ICPD "Programme of Action," however, have already given it significant support. Others, for principled reasons, oppose some of the proposals. What impressed conferees at the Genval meeting was the surprising number of themes on which there was convergence in the assessments of the ICPD recommendations and religious responses to them.

Findings

1. On freedom of religion and conscience, and the role of religious communities in international debate on public policy.

Any talk about controversial international issues must include discussion of freedom of religion and even freedom from religion and religions. The ICPD, of course, cannot and should not reach into particular religious communities and seek to impose its will on the consciences of individual believers. But international bodies could not achieve anything of importance for the larger world community if they were never permitted to challenge the religious outlook of one or more faiths or never allowed to develop programs that one or another of the religious communities might oppose. Religious groups themselves must respect the beliefs and values of others, because no single faith may claim final moral authority in international discourse.

Such assertions call for prudence on the part of religious communities, including the largest ones, which can exert great influence upon national and international policies. They, of course, must be free to guide members of their own communions on issues and practices deemed essential to the integrity of their religious traditions and to try to persuade others of their positions in public debate. So also they must be ready for criticism when some of their practices offend others or seem to those others to work against the sacredness of life and human dignity.

The corollary of religions' freedom to propagate and practice a faith is the freedom of others not to be bound by their strictures. On almost any vital topic addressed by the ICPD there will be some religious dissent. Although it is important for strategic and humane reasons to take such dissent into consideration, the UN and its member nations will also, at times, address issues in terms uncongenial to one or another religion. Some faith communities may perceive themselves as the possessors of all truth and the custodians of all virtues in both religious and political spheres. But because we exist in time and space and share the earth with others, one community's claim to truth and virtue is not likely to be accepted by all others. In the political realm, the voices of different faiths and moralities need to be heard. But the voice of any single faith should not carry so much weight as to stifle debate or paralyze action on the international agenda.

2. On the population and development crisis.

People of faith across the boundaries of particular traditions, because of their religiously inspired concern for the health and well-being of persons and the environment, agree that a population and development crisis exists. They may not fully concur on how the issues interrelate or on what means for addressing the crisis can be religiously endorsed. But when confronted with evidence about current conditions and future implications of the crisis, most recognize complex linkages among the issues and an urgent need for the international community to address these factors simultaneously.

3. On development, consumption, and the maldistribution of resources.

In light of the history of what has occurred in the name of *development*, the term is sometimes suspect. Because its usage is a given in ICPD documents and discussions, however, many religious communities would urge the adoption of a comprehensive definition of *development* that unequivocally embraces a broad understanding of human nature and need, a sensitivity to cultural distinctiveness, a respect for environmental integrity, and a deep regard for justice and equity.

Most religious traditions have much to say about caring for resources and using them responsibly, about valuing human work and planning and about providing equitable access to both the means and fruits of development. They speak out against waste and extravagance, greed and self-preoccupation, lack of generosity and unwillingness to share, and, most of all, against personal and structural injustice that prevents sustainable development and impedes access to the means of supporting not just basic survival but also human well-being and environmental integrity.

While religions vary in their assessment of the value of wealth, over-consumption by individuals and within whole societies is seen as problematic in virtually all the faith communities and their classic religious texts. This concern is intensified when over-consumption results in pollution of the environment and depletion of irreplaceable resources.

No consensus exists on how over-consumption by the affluent may confound development in poorer countries. But virtually every religion instructs adherents to attend to the needs of the poor, and much religious energy has been directed to the question of how the poor and the hungry are to be treated by individuals and in societies. They are usually seen to be especially deserving of attention.

In a global society, these moral insights apply far beyond the boundaries of individual religious communities or nations. The means of achieving the eradication of poverty and hunger, the enhancement of human well-being, and the protection of the environment will need to be broad-based and multifaceted and will require the cooperation of religious and civil communities.

4. On humans and the natural environment.

God-centered religions speak of the natural environment as the work of a creator or creators and therefore as sacred. Nontheistic faiths also affirm the sanctity of the natural environment. Almost all the texts that witness to the founding of faith communities include injunctions to stand in awe of nature, to respect it, to be good stewards of its resources, and to share them with some sense of fairness.

Some religious texts speak of human dominion over the natural world and have been invoked, at times, to sanction the exploitation of nature for human use. More recently, many faith traditions have begun to emphasize that human well-being depends upon the quality of air, earth, and water in one's community and around the globe. Moreover, a sense of the intrinsic value of nature, long prevalent in some traditions, now grows in other religious faiths as well. The human may be sacred but is not alone sacred.

5. On the role of women in issues of population and development.

The witness of religions to the equal dignity, rights, and freedom of women and men has not been uniform or unambiguous. Many ancient texts and modern interpretations reflect cultures of male dominance and are construed as divine authorization for the subjugation of women. When religious communities meet, they often speak self-critically about past abuses and continuing resistance to acknowledging the full humanity and moral agency of women. But almost without exception the world's religions are faced with the challenges of not only modern movements but also those sacred texts and traditions, often obscured in the past, that either explicitly or implicitly affirm the dignity, moral agency, and equivalent roles of women and men. How to advance women's rights and roles is debated among and within religious communities and cultures. That these must be advanced—or in some cases retrieved from what had originally been set forth—is not at issue among them.

The claims for the full human dignity of women in the private and public as well as sacred and secular spheres of life, therefore, will likely find increasing support among and within religious communities, even when the official position of a religious tradition is at variance with the views held within the larger body. And while women in different parts of the world may reflect the distinctive character of their cultures and faiths, the fundamental and universal rights of women as human beings can be expected to receive support from a broadening spectrum of religious communities.

These basic rights require the participation of women in the formulation and implementation of policy, particularly in those areas where their lives are most directly affected, and especially in such vital areas as reproductive health, population growth, development, and the environment. Women individually are also understood to have the freedom to exercise conscience in matters that have an impact on their survival, health, well-being, and destiny. For this freedom to be meaningful, women need access to education, the resources for reproductive health, and opportunities for personal development and socio-economic advancement.

6. On religious communities' valuing of families in their various forms.

Religions tend to hold relatively conservative views on families, although over a long period of history many religious communities have embraced a variety of family forms. Some models of family have been so repressive, often with religious sanction, that they have drawn criticism from within and outside faith traditions. Others, e.g., religious inspiration, have experimented with variant forms of marriage and family life. While religious traditions in different times and cultures have sanctioned various family structures, they have consistently viewed families as hallowed institutions that should be protected and enhanced.

Religious communities regard the family as the chief place in which the religious and cultural values of communities and civilizations are transmitted from generation to generation, thus providing stability and moral cohesion to a people. Increasingly today religions too are recognizing that abuse and human rights violations within families cannot be condoned or justified.

Families are also important for faith communities because children need sustained nurture in a reliable context that some form of family helps assure. Children require material and human resources for their survival and development, and the capacity to provide for these essentials should be a factor in family planning. Issues concerning these needs are as urgent to faith communities as are debates over reproductive rights and policies. In such infinite circles of life, boys and girls can grow toward adulthood with a developing sense of moral rootedness and responsibility.

7. On the special circumstances of adolescents.

Different religious traditions have different understandings of the roles, needs, and expectations of adolescents, as do different cultures. Thus it is difficult for international bodies to offer proposals or programs that are applicable and appropriate to adolescents everywhere. Still, because adolescents are capable of sexual activity, which can have enduring consequences for their own health and well-being and that of their offspring, as well as for societal stability, concern about ill-informed or irresponsible sexual behavior is legitimate. Most religious traditions tend to address this concern by urging sexual restraint on the part of unmarried adolescents.

Neither faith communities nor civil societies, however, can expect responsible sexual behavior on the part of adolescents without providing information, moral guidance, and education. Faith traditions can be instrumental in helping adolescents refrain from premature and irresponsible sexual behavior if the religious communities accept their obligation to engage in effective religious and moral formation. The occasion of the ICPD provides a timely opportunity for religious communities to develop training programs for parents and teachers of adolescents reflecting their distinctive teachings on sexuality and responsible sexual behavior.

8. On contraception as an instrument of reproductive and public health, family planning, and population stabilization.

Although a few major religious groups or movements steadfastly oppose "artificial" contraception, almost all of the world's religions endorse contraception as a means of improving reproductive and public health, promoting family planning and responsible parenthood, and contributing to population stabilization. Contraception is also seen as an effective way of reducing the incidence of abortion. In those few instances in which a religious body condemns artificial contraception in its official teachings, evidence indicates that the teaching is widely disregarded by followers, even those who are loyal in other areas of religious discipline. This suggests that many people of faith today see no important moral distinction between natural and artificial methods of contraception.

Religious communities, however, object to any form of contraception that is coercive or is forced by governments on citizens, because such action is a profound violation of human freedom and moral agency in a fundamental area of human life.

Voluntary contraception, on the other hand, is seen as a way in which those cherished values can be acted upon. Because of a strong commitment to the physical, mental, and spiritual health of human persons and the well-being of families, as well as a profound respect for the exercise of personal conscience, most religious communities view contraception as a responsible option, even as a religiously motivated activity.

9. On abortion.

While abortion is universally treated as a serious moral and religious concern, it is treated differently among and within religious communities. Most religious traditions do not forbid abortion altogether, yet some limit the conditions under which it may be permitted. Others understand it as a matter that is to be left to the discretion of the individual in conformity with the dictates of personal conscience.

The rationales for taking one position or another on abortion among and within religious groups include a commitment to the sanctity of human life even in its earliest embryonic stages; an interest in the current and future health and well-being of both the mother and the embryo or fetus; a respect for the right of a woman to act as a full moral agent; and a concern that the state not interfere in personal matters of conscience. Although some traditions and people of faith ground their stance in one of these rationales to the exclusion of the rest, many others develop positions that draw on several of these perspectives, resulting in a great variety of assessments concerning when and under which circumstances abortion is morally justified.

Whatever their stance on abortion, religious communities cannot disregard the fact that it occurs and that, in places where abortion is illegal or heavily restricted, it often poses risk to the life and health of the woman. Decriminalization of abortion, therefore, is a minimal response to this reality and a reasonable means of protecting the life and health of women at risk.

Given the moral concern about abortion and the range of stances toward it taken by religious communities, the view of any particular religious tradition should not be imposed on others.

10. On sex education, particularly in the context of sexually transmitted diseases and the AIDS crisis.

Many faith communities in diverse cultures have been so protective of traditional ways and so defensive about sex education under the auspices of "others"—secular agencies or other religions—that they have neglected to offer such education on their own and to their own. As communication between generations breaks down and a variety of forces assault the integrity of families, young people and adults alike are often without moral guidance on issues of sexuality, sexually transmitted diseases (including AIDS), and responsible parenthood. Religious communities must take up anew their responsibility to act as educators on such vital issues. They should also counter those elements of mass media and global popular culture that undermine a full appreciation of human sexuality and responsible sexual conduct.

In cultures where one religion predominates, special attention should be given to acknowledging and respecting the moral convictions of those who hold minority views on sex education. A single system should not be imposed.

11. On human migration in a time of unsettlement.

The religious communities have much to say about the movement of human populations—whether in pilgrimage, hajj, exodus, exile, the search for utopias and promised lands, wanderings, explorations, or for refuge. Religious texts address the plight of those left abandoned, homeless, or impoverished as a result of war, exploitation, and natural disaster. Those without homes deserve special attention and compassion.

The twentieth century has seen unprecedented numbers of both voluntary migrants and refugees who have been forced from their homes; continuing warfare and material deprivation will displace even more. Unfortunately, religious warrants for "ethnic cleansing" and tribal aggression have been invoked all too often. It should be noted, however, that faith communities have often provided protection and assistance to forced migrants, political refugees, and victims of war and famine.

Because many religious traditions promote special treatment for the homeless and for strangers, religious communities recognize urgent reasons to connect the themes of dislocation and resettlement with those of population and development.

Afterword

Participants in the consultation reaffirmed the vital interest of their religious communities in the interrelated issues of population and development, noting the impact that policy decisions have on the diverse peoples of the earth and the environment, along with the constructive contributions religious communities can make in forming and implementing public policy in these areas. Reflecting on their experience together, they acknowledged that exploring these issues in a multifaith and multicultural context enriched understanding and the possibility of cooperation. They recommended that similar discussions continue at the international, national, and local levels on the vital and interrelated issues of population and development.

A Catholic Perspective

David M. Byers*

We were sitting in the lounge at Fordyce House after a long day of discussing global population at the conference of the Institute for Theological Encounter with Science and Technology (ITEST) held in St. Louis. Monsignor Diarmuid Martin leaned across the table and asked Dr. Alene Gelbard, "Could we have had this conversation five years ago?" Dr. Gelbard thought for a moment and then responded, "I don't think so. Attitudes have softened."

It was a hopeful moment for the future of religion–science collaboration on the topics of consumption, global population, and the environment. Monsignor Martin headed the Vatican delegation at the UN Conference on Population and Development in Cairo; Dr. Gelbard directs the international programs of the Washington-based Population Reference Bureau. I was privileged to be the third speaker at the conference, talking about ongoing dialogue between the scientific and religious communities.

Dr. Gelbard's point was that the population experts of the developed world have begun to have second thoughts about the utilitarian goal that mesmerized many of them in the 1970s and 1980s: reaching numerical population goals without worrying much about means. The Cairo Conference marked a substantive turning point in the discussion of population, and therefore about ecology in general. A recent article in *Science* noted several "conceptual shifts," among them movement "from a single-minded concern with population growth and the need

*David M. Byers is executive director of the National Conference of Catholic Bishops' Committee on Home Missions and the Committee on Science and Values.

for population control to a framework for approaching population-related issues that considers the interactions of population, poverty, and the patterns of consumption and production; . . . from a technological approach to fertility control to comprehensive reproductive health services joined with broader social action in other areas, including education for women [and] legislation to prevent discrimination against women and girls . . . [and] from narrowly defined family planning programs that aim to reduce fertility to an emphasis on health, empowerment and the right of individuals to determine the number, spacing and timing of children."[1]

Dr. Gelbard and Monsignor Martin's conversation was a telling vignette, in that the religious community is partly, perhaps even largely, responsible for these shifts. The Vatican delegation at Cairo demanded that the conference take seriously such truths as God's love for each one of us, respect for life and for the dignity of the human person as *imago dei*, gender bias in having and raising children, the equality of women, the importance of the family, the need for justice in the distribution of material goods among peoples and nations, and responsibility for future generations. From the perspective of the Catholic Church, these contributions are important aspects of "the truth about man," to use a favorite phrase of Pope John Paul II. In a 1994 address to the Pontifical Academy of Sciences, the Holy Father asserted: "In his mystery, man goes beyond the sum of his biological characteristics. He is a fundamental unit, in which the biological cannot be separated from the spiritual, family and social dimensions without incurring the serious risk of suppressing the person's very nature and making him a mere object of analysis."[2]

The church's integral view of humanity leads it to dismiss reductionism out of hand and to reject any solutions to problems of population, consumption, or the environment that are mechanistic, simplistic, or, above all, coercive. In this it does no more than ground the exalted yet compassionate view of the human person—in Hamlet's phrase, "this quintessence of dust" who is also "the beauty of the world" (II, ii, 325–26)—that has inspired all great art, music, and literature. When true to itself, religion civilizes, teaching us to love one another as we are while pursuing a vision of what can be.

The Catholic Church is particularly well situated to add this civilizing note because of its global reach and venerable moral tradition. For example, the church's varied membership forces it to see the use of material goods from complementary perspectives whose juxtaposition should yield wisdom. Its members view consumption with Southern as well as Northern eyes; they are Lazarus as well as the rich man.

Catholic moral tradition, meanwhile, by painting a tormented rich man confronting Lazarus in Abraham's bosom, offers permanently valid guidance. John Paul II offered a most striking example of such guidance in his October 19, 1995 address to the United Nations, in which he reminded the countries of the world that they form a "family of nations." In his words: "The idea of 'family' evokes something more than simple functional relationships or a mere converg⌐ of interests. The family is by nature a community based on mut   ⌐st, mutual support and sincere respect. In an authentic family   strong do not dominate; instead, the weaker members, because of their very weakness, are all the more welcomed and served."[3] To paraphrase Theodore Roosevelt, the papacy is the world's bulliest pulpit from which to preach about the common good, the things that affect us all.

The environmental crisis not only gives the church a chance to contribute, it also challenges the church in certain ways. Some would suggest that the greatest challenge is to reform Catholic teaching on family planning, especially artificial birth control. This, it seems to me, is rather unsubtle. No doubt the church has a long way to go in presenting this teaching in a form that makes sense in the modern world, even to its own members. This effort may or may not bear fruit as the debate matures. However, criticisms of Catholic teaching are superficial unless they take into account the physical and moral context of the population question. The Vatican position paper on the 1992 Earth Summit notes, "The relationship of development and the environment to population growth is complex and often tenuous,"[4] a point that the Cairo Conference, for all its enthusiasm for contraceptive practices, generally supported. Discussion of contraception also often ignores cultural differences and the very troublesome problem of enforcement. Who makes the decisions? Who takes the orders? If population control is not approached with a healthy respect for human rights, it runs the risk of becoming the latest and most invasive exercise in first-world colonialism.

Turning to the national scene, one must admit that the Catholic Church in the United States does not have a long tradition of attention to the environment. However, this is changing fast in response to the Holy Father's World Day of Peace message for 1990, "Peace with God the Creator, Peace with all Creation." In 1991, the National Conference of Catholic Bishops (NCCB) issued *Renewing the Earth: An Invitation to Reflection and Action on Environment in Light of Catholic Social Teaching*. This pastoral statement covers the issues of this book and more, with sections on a Christian understanding of stewardship and on the preferential option for the poor. The bishops later set up the

Environmental Justice Program to "help Catholics in parishes, diocesan institutions and organizations, and national organizations to take seriously the moral obligation to care for God's gift of creation."

This new program carries out a range of activities, including offering small grants in support of parish and diocesan initiatives that can serve as models for others. Early last year, the program issued *Renewing the Face of the Earth: A Resource for Parishes,* which packages the text of the pastoral with scriptural and magisterial reflections on stewardship, environmental justice, and other topics; with suggestions for prayer and worship; with recommendations for integrating environmental concerns into parish life; and with thoughts on what individual Catholics can do to help the environment. A second resource packet recently appeared. It contains a prayer service and homily aids on environmental concerns, together with articles on ecological spirituality, environmental hazards, and the problem of consumption.

One other NCCB effort, the Bishops' Committee on Science and Human Values, which I staff, exists to carry on dialogue with the American scientific community. The committee aims to inform the bishops about advances in science and technology, and to bring Catholic moral thought to bear on these advances. In 1993 and 1994, the topic of our annual sessions was global population/global resources. While the participants spent a good bit of time discussing scientists' stereotypes of bishops and vice versa, we were able to define some principles we held in common. The scope of agreement was severely limited; the bishops and scientists did not define the terms of these principles with any precision or outline how they might be applied. Nevertheless, the Committee on Science and Human Values was able to report substantive results to the U.S. Catholic hierarchy, increasing the bishops' comfort with demographics and their knowledge of its pastoral relevance. As a side effect, the dialogue provided material for a popular brochure entitled *Science and the Catholic Church,* which declares unilateral withdrawal from the mythical war between religion and science.

I said earlier that I do not consider artificial birth control the greatest challenge that the environmental crisis presents the church. The greatest challenge is the same one that unjust war and hunger and cruelty and adultery and oppression and egomania and any other form of human sinfulness presents to us: the need to evangelize. Catholics are not only in the civilizing business, as important as that is. We are also in the saving business, preaching Jesus Christ and him crucified to the man next door, to the nations, even to ourselves. Our greatest challenge, then, is also our greatest contribution. The church must approach the environmental crisis proclaiming that we are creatures in the midst of creation, not independent agents with adding machines. It must approach the environmental crisis with love, with that gentle spirit which Paul recommends to Timothy (I Tm 6:12). I recently saw an interview with Cardinal Joseph Bernardin of Chicago who was recovering from a very serious operation for cancer. "We spend so much time on things that are not so important," he said. "What does make a difference is whether what is done, whether what is said, really contributes to the well-being of people, individually or collectively."[5]

Notes

1. Claudia Garcia-Moreno and Tomris Turmen, "International Perspectives on Women's Reproductive Health," *Science* (August 11, 1995), pp. 790–91.
2. John Paul II, "Address to Pontifical Academy of Sciences," *L'Osservatore Romano* (English Version), November 9, 1994, pp. 9.
3. John Paul II, "Address to the United Nations," *Origins* (October 19, 1995), pp. 294–98.
4. Vatican City State, "Earth Summit: Environment, Development and Population," *Origins* (June 11, 1992), pp. 70–72.
5. "Cardinal Bernardin Considers His Illness 'A Great Blessing,'" *Catholic News Service*, October 26, 1995, pp. 9–10.

Foundations for a Jewish Ethic Regarding Consumption

Michal Fox Smart*

Among the privileged majority, many North Americans feel a sense of embarrassment or shame that we consume such a disproportionately large percentage of the world's resources. For instance, although we constitute merely 6 percent of the human population, we consume approximately 40 percent of the energy produced.

The costs of our level of consumption are significant, and we are increasingly familiar with its ill effects.[1] Amid the effort to acquire wealth, half of all Americans feel that they don't spend enough time with their spouses, children, and friends. Thirty percent of all adults report experiencing high stress nearly every day. Moreover, we are depleting the planet's resources at an alarming rate; denying habitat and sustenance to countless other creatures; reducing the prospects of future generations; and widening the gulf between rich and poor.

Given the deep roots and pervasiveness of our consumerist culture, societal transformation can seem an overwhelming task. One can begin, however, by recognizing that the set of problems commonly referred to as the "environmental crisis" are actually only symptoms of what is essentially a crisis of values. Values education is, of course, a difficult matter in the public sphere. The question immediately arises as to whose values to teach, and in what language and form. It is for this reason, among others, that religious leaders and communities have a

*Michal Fox Smart is a Jewish environmental educator. She is currently a full-time mother in Hartford, Connecticut.

137

unique and critical role to play in addressing this crisis at its deepest level. In formulating an ethic regarding consumption, several aspects of Jewish tradition will be fruitful to explore.

The Need for Balance

As a whole, Jewish tradition is virtually unswerving in its validation of consumption for reasons beyond sheer physical sustenance, including celebration and religious ritual. The this-worldly orientation that typifies mainstream Judaism reveals few ascetic or "anti-materialistic" tendencies, and by and large seeks to sanctify rather than to denigrate biological acts such as eating, procreation, and bodily pleasure. In the post-Temple period, for instance, the everyday act of consuming food at one's table is likened to performing ritual sacrifices upon the altar of the Holy Temple. In a similar vein, Judaism provides blessings for wearing new clothes and for eating a fruit for the first time each season. There is a pervasive notion that certain kinds of consumption do not denigrate Creation nor the consumer, but instead enable one to experience the holy within the everyday. One sage in fact reports that on the day of judgment, one will have to account for everything he could have enjoyed from this world but did not.[2]

At the same time, however, many voices within the tradition view excessive consumption and the hoarding of goods as vices. "Envy, self-indulgence, and prestige-seeking drive a person out of the world," declares Pirqei Avot.[3] "The more possessions, the more worry," another sage observes.[4] Nahmanides, a medieval biblical commentator, explains the biblical exhortation "and you shall be holy"[5] as admonishing one who consumes permitted substances in excess, thereby becoming a "knave within the bounds of the law."[6] Thus, we are exhorted on the one hand to participate fully in the drama of Creation and to enjoy the gifts that God has provided, and on the other hand to avoid gluttony. The challenge, apparently, is to consume discerningly, and the task remains to define principles whereby a proper balance may be achieved.

Foundational Ideas: The Earth Is the Lord's

"In the beginning," the Torah tells us, "God created Heaven and Earth."[7] From the perspective of Jewish tradition, the world is not a fortuitous cosmic accident but rather a purposive creation of God. Judaism's starting place is not that everything is permissible, but rather that everything is holy. For this reason, the rabbis said that one who destroys objects in anger should be regarded as an idolater.[8]

Moreover, the fundamental belief in God as Creator is traditionally understood to imply that the world does not belong to human beings to exploit at will but rather still belongs to God. In the language of Psalm 24, "The earth is the Lord's and all which fills it, the world and all its inhabitants."[9] Jewish tradition as a whole rests upon the premise that we do not have unrestricted use of creation. Countless laws serve to mediate Jews' interaction with the world, detailing which actions and parts of God's world are allowable to us and when, and which are forbidden.[10] Indeed, it is precisely Jewish law's restriction of one's autonomous will that runs so counter to modern sensibilities.

Within Jewish tradition, this central tenet of divine ownership is manifested in daily practice. For instance, because the earth and all which fills it belong to God, the Talmud states: "It is forbidden to enjoy [anything] from this world without a blessing."[11] Rav Chanina Bar Papa, a talmudic judge, said: "Anyone who enjoys [something] from this world without [reciting] a blessing, it is as if he steals from the Holy One, Blessed be He. . . ."[12] Thus, blessings continually remind us, each of the many times each day that we take of its bounty, that the world belongs to God and that our use of its richness must be responsible.

Social Justice

God's perceived ownership of the world is further manifested in Judaism's bold vision of social justice. *Zedaqah*, the Hebrew word for giving to those in need, does not mean charity. It means justice. For all we possess is ultimately a gift of God. The wealthy are no more deserving of wealth than the poor, nor do they own their wealth in an ultimate sense. It is not generosity of spirit, therefore, but righteousness that demands that resources be redistributed and that these basic needs of all people be met.

Extensive *Halakhic* (Jewish legal) material governs the distribution of resources within Jewish society. It will serve our discussion to highlight only a few of its central characteristics and practices.[13] With overwhelming consistency, mainstream Judaism recognizes a basic level of economic sustenance as necessary and good. A "bottom line" position is asserted in the Mishnah[14]: "Where there is no flour [i.e., material basis for sustenance], there is no Torah. Where there is no Torah, there is no flour."[15]

Throughout the ages, Jewish communities across the globe have been characterized by a vast range of social institutions designed to provide for those in need. Support of those institutions was not voluntary; taxes were enforced, and communal leaders even reserved the right to confiscate the assets of a Jew who reneged on his charitable responsibilities. Thus, the right to acquire wealth, like the right to own private property, has been upheld by Judaism yet tempered by an overriding concern for justice. Still today, *Halakhah* dictates that one give one tenth of his or her income to those in need, and that even those who receive public assistance devote a portion of their income to others in still greater need.

The laws of zedaqah are nuanced. They demand not that all community members be accorded the same level of wealth, but rather that basic needs such as clothing, food, and shelter be provided for all. Maimonides, a medieval scholar, reasoned that the highest form of charity is to provide the means for a person to earn an independent living.[16] Interestingly, the obligation to donate at least one tenth of one's income is balanced by a limit: One is not to give away more than one fifth of what he or she has. Jewish law seemingly recognizes that an ethic of living simply and/or giving to others can be taken too far, to the point where oneself or one's own family becomes impoverished.

These laws, while important in themselves, can also serve as both a practical model for other communities and a basis for a broader ethic within the Jewish community. For given the explosion of the human population and the rise of the global marketplace, the same concern for social justice and equity within contemporary Jewish society that underlies these laws urges us to provide other nations and future generations with the resources they need for a full life, as well. We can do that only by reducing our current consumption of energy and material goods.

Ecological Justice and Care for Creation

Witnessing the habitat destruction and species extinction to which our consumption patterns contribute, we must today expand our thinking to consider not only social justice among people, but also ecological justice—meaning just relations between human beings and other creatures. There is little basis within Jewish tradition for constructing a rights-based argument for other creatures or inanimate elements, as Judaism as a whole is based more on duties than on rights. Nonetheless, scripture clearly conveys that God values nonhuman life and cares for other creatures. One need look no further than Genesis 1 to see that God considers creation—flora and fauna, energy, and inert matter—to be good. The blessing to "be fruitful and multiply" is given not only to humans, but also to all species of fish and birds.[17] Several psalms depict other species sharing in the bounty of creation. The last chapters of the book of Job exult in a stunning panorama of biodiversity. An increasingly well-known *midrash* declares, "Even those things you deem to be superfluous, such as mosquitoes, gnats, and flies, even they are included in the scheme of Creation, as it is written: 'And the Heaven and the Earth were completed, and all their hosts.'"[18] These texts are complemented by biblical passages that use covenantal language to describe God's relationship with the land and other creatures,[19] as well as by laws that protect the integrity of created kinds from contamination[20] and, according to some interpreters, in fact prohibit the extinction of an animal species.[21]

The import of these texts is clear: Divine care extends beyond human beings to the whole created world. Not only the flourishing of human culture but also the thriving of plants and animals in their natural habitats is a divine command and blessing. Accordingly, in the traditional "Grace" after "meals," Jews praise God for providing nourishment not only to human beings, but also for sustaining all life.

This should foster an attitude of humility on the part of human beings when influencing the fate of other creatures. Too often we reduce to cost-benefit analysis creatures whose significance cannot be measured in regard to their utility to this generation of humans—indeed, whose ultimate purpose lies shrouded in the mystery of Creation itself. ("Where were you when I laid the foundations of the earth?" God asks Job, when he presumes to know just how things should work.) A religious value of caring for Creation, with a derivative notion of ecological justice, cautions against patterns of consumption that result in the destruction of natural habitats and that threaten fellow creatures with extinction.

The Connection between Ecological and Social Justice

Furthermore, Judaism sees an integral connection between social (i.e., human–human) justice and ecological (i.e., human–rest of creation) justice. Time and again, specific religious observances contain both of these elements, interwoven both practically and philosophically. For example, *Shemitah*, the Sabbatical year, is primarily defined as a Sabbath for the land: "And the Lord spoke to Moses at Mt. Sinai saying, 'Speak to the children of Israel and say to them: When you come into the land which I give to you, the land shall observe a Sabbath to the Lord.'"[22]As a consequence of the earth's Sabbath, privately owned land is legally deemed ownerless, and standard cultivation of crops is forbidden. A striking aspect of Shemitah observance—*biyur*—necessitates that once a given food is no longer available to wild animals in the fields, this same item must be removed from human domiciles and declared ownerless.[23] Thus, observance of Shemitah serves to deflate the chasm between human and wild animal, between the natural world and human society.

Significantly, this "Shemitah of land" is accompanied by a "Shemitah of money," in which all outstanding monetary debts among Jews are forgiven in an effort to eliminate poverty within the nation. Shemitah thus addresses both inequity in the social realm and inequity between humans and the rest of creation.

Particularly at a time when the vast majority of toxic dumps, landfills, trash incinerators, and the like are located in poor and often predominantly minority neighborhoods, and we are confronted by a host of complex issues regarding population control and "third-world" development, the connection between these two realms of concern is profound.

The Prohibition against Wanton Destruction

In articles over the last twenty years and in classrooms around the country, much attention has been paid to a set of laws called *bal tashchit*, literally "do not destroy." These laws are based in a prohibition in Deuteronomy 20:19, that when laying siege to a city in time of war, one cannot cut down fruit trees in order to make a bulwark to aid in the siege. The rabbis later extrapolated from this unique circumstance a host of prohibitions that as a whole forbid the wanton destruction of anything of utility, whether natural or human-made. Because this biblical law explicitly addresses the cutting down of trees, and because rabbinic law at first glance expands this law to mandate the conservation of virtually everything of value that exists, *bal tashchit* is frequently embraced as a foundation for a Jewish environmental ethic.

In fact, the laws of bal tashchit are frequently misunderstood. The biblical verses on which they are based are oblique, and a variety of possible translations yields significantly different interpretations of their import.[24] The range of medieval commentary on these verses is in fact fascinating and encompasses a wide range of philosophical approaches to the matter of consumption. For instance, Rashi, the great medieval commentator, interprets these verses to teach that because trees are not human beings, they should not be drawn into human conflict; that they have, in essence, moral standing independent of human needs and the human drama by which they are surrounded. In contrast, other commentators understand the fruit trees to be protected because human beings are dependent on them as resources for food. According to this reading, one should preserve them as an act of prudent self-interest and, moreover, faith in the future.[25] Study of these verses therefore draws the reader into discussion of the relationship between human beings and trees of the field and the moral and practical concerns that should govern that relationship.

Rabbinic law expanded the scope of bal tashchit by extending its legal purview from fruit trees to virtually any material object. For instance, the Talmud states: "Whoever breaks vessels, or tears garments, or destroys a building, or clogs a well, or does away with food in a destructive manner, violates the negative commandment of bal tashchit."[26] Yet the rabbis also greatly eclipsed the discussion philosophically by defining the principle of bal tashchit in strictly economic and utilitarian terms. Maimonides summarizes the talmudic discussion: "It [a fruit tree] may be cut down, however, if it causes damage to other trees or to a field belonging to another man or if its value for other purposes is greater. The Law forbids only wanton destruction."[27] Thus, according to rabbinic law, one is permitted to cut down fruit trees if the resulting lumber is more economically valuable than the harvest of its fruits. Likewise, trees that are in any way a nuisance may be cut down. As the laws of bal tashchit were developed in Jewish tradition, then, the value they ascribe to things—whether living trees or human artifacts—is strictly instrumental. They prohibit only the *wanton* destruction of resources; destruction is allowed if human benefit is perceived to be derived thereby.

The following talmudic passage is illustrative. "R. Hisda said: When one can eat barley bread [which was presumably inexpensive to make] and eats wheat bread [apparently more costly], he violates bal tashchit. R. Papa said: If one can drink beer and drinks wine, he is also in violation of this verse." "However," the Talmud then continues in the voice of the redactor, "this is incorrect. For the prohibition 'not to destroy' oneself is more important."[28] We have here, then, the suggestion of an ethic of extreme frugality, followed by the conclusion that some higher level of consumption—at least to the point of indulging in wheat bread and wine—is life enhancing and therefore justified. Thus, the law of bal tashchit does not define a point at which an act that gratifies human desires should nonetheless be censured.

While it is not the foundational tenet that many propose it to be, the law of bal tashchit does contribute in a limited way to the articulation of a Jewish ethic regarding consumption. First, many of us today are so wasteful that certain aspects of our consumption patterns may indeed constitute wanton destruction and therefore violate the law, even in its narrow definition. For instance, bal tashchit would seem to prohibit the purchase of clothing, toys, or equipment that will never be used. Returning to the biblical text, bal tashchit also cautions against sacrificing the long-term usefulness of a resource (e.g., fruit trees) for immediate, short-term use and gratification (e.g., bulwarks during a siege). While the law does not specify the time frame that such an analysis must encompass, at least one traditional reading of the biblical prohibition understands bal tashchit to mandate the preservation of resources for future human need. Understood in this way, the principle of bal tashchit may temper the greed of any one generation in light of the ongoing needs of generations to come. Furthermore, rabbinic law invokes the principle of bal tashchit to mandate efficiency in consuming resources when the immediate benefit to people is unaffected. "Rabbi Zutra said: He who covers an oil lamp, or uncovers a naphtha lamp, infringes the prohibition of bal tashchit, since these acts cause the lamp to burn with unnecessary speed."[29]

Looking Inward, Looking Ahead

In the final analysis, however, intellectual exploration of the tradition is not sufficient to meet the challenges that confront us. For ultimately the factors that drive one to consume are largely attitudinal and reflect not only one's thinking, but also one's wealth in nonmaterial sources of satisfaction. If we, as a society, are to wean ourselves from our unsatisfying addiction to material consumption, we will need, each in our own way, to develop our inner resources and achieve a sense of spiritual fulfillment and well-being. A person who seeks to satisfy his existential needs through financial security will never have enough. "Who is rich?" our sages asked. Then they answered, "He who is satisfied with his portion."[30] And who is it, we must further ask today, who will be capable of being satisfied with her portion? I suggest it is someone whose spiritual needs, above and beyond her physical needs, are sufficiently met— one who has infused her life with a sense of purpose and belonging; of participation in a story larger than herself, which is ennobling, and who has a solid and healthy connection to herself, to family, to community, and to God.

At its best, religious education has the capacity to shape the human personality and to inculcate sensitivities and an inner awareness that are critical resources in our age. Sadly, religious education infrequently plays such a formative role in the lives of contemporary Jews, and herein lies perhaps the greatest challenge we face. In seeking to restore the balance in our consumption of goods and resources, we do not need religion to stretch beyond its natural borders to address something outside of itself, e.g., "the environment." Rather, we need a process of religion coming into being—coming to claim and utilize its full power to instruct and inspire, to engage in thoughtful and meaningful ways with issues of contemporary concern, and to inform the moral life of its community.

Jewish engagement with issues of consumption and stewardship is a slow and complex undertaking—one as rich in challenges as it is in insights revealed. Thankfully, however, we are indeed moving forward in an effort to articulate a Jewish environmental ethic and to understand that we must be responsible members of a community that ultimately includes all peoples, and God's world as a whole. Only by striving to envision and create an equitable society, with both social and ecological justice considered, can we fulfill our most essential mandate as Jews: to bear witness to God as Creator of heaven and earth, and to call others to live in a manner that reflects this most awesome of truths.

Notes

1. See Alan Durning, *How Much Is Enough?* (New York: W.W. Norton, 1992).
2. Y. Qidushin 4:12, *Jerusalem Talmud.*
3. Pirqei Avot 4:21, *Mishnah.*
4. Pirqei Avot 2:8.
5. Leviticus 19:2, *The Holy Bible.*
6. Nahmanides, ad loc.
7. Genesis 1:1, *The Holy Bible.*
8. See *Babylonian Talmud,* B. Shabbat 105b: "One who tears his clothes in anger, or one who breaks his tools in anger, or one who wastes his money in anger you should consider as an idolater."
9. Psalm 24:1, *The Holy Bible.*
10. The story of the Garden of Eden can itself be understood as a paradigmatic tale in which human beings express their desire to be Godlike by refusing to accept limitation on their consumption of the world.
11. *Babylonian Talmud,* B. Berakhot 35a.
12. *Babylonian Talmud,* B. Berakhot 35b.
13. For a fuller discussion, see Meir Tamari, *With All Your Possessions: Jewish Ethics and Economic Life* (New York: Free Press, 1987). I am indebted to Rabbi Eliezer Diamond in the following discussion.
14. The *Mishnah* consists of six orders of the Oral Law, as redacted in the third century by Judah ha-Nasi.
15. Pirqei Avot 3:21.
16. Maimonides, *Mishneh Torah, Hilkhot Matanot Aniyim.*
17. Genesis 1:22, *The Holy Bible.*

18. Leviticus Rabbah 22.
19. See, for example, Genesis 9:9–10, *The Holy Bible*: "Behold, I establish my covenant with you, and with your seed after you, and with every living creature that is with you of the birds, of the cattle, and of every beast of the earth with you, from all that came out of the ark, to every living being of the earth."
20. See Leviticus 19:19, Deuteronomy 22:9–11, *The Holy Bible*.
21. See Deuteronomy 22:6–7, Leviticus 22:28, *The Holy Bible*; and Nahmanides, ad loc.
22. Leviticus 25:1–2, *The Holy Bible*.
23. M. Shevi'it 7:1.
24. See Deuteronomy 20:19–20, *The Holy Bible*: "When in your war against a city you have to besiege it a long time in order to capture it, you must not destroy its trees, wielding the axe against them. You may eat of them, but you must not cut them down. Are trees of the field human to withdraw before you under siege? Only trees which you know do not yield food may be destroyed; you may cut them down for constructing siegeworks against the city that is waging war on you, until it has been reduced." (New Jewish Publication Society Translation).
25. See for example, the commentary of Ibn Ezra, ad loc.
26. *Babylonian Talmud,* B. Qidushin 32a.
27. Maimonides, *Mishneh Torah,* Laws of Kings and Wars 6:8.
28. *Babylonian Talmud,* B. Shabbat 140b.
29. *Babylonian Talmud,* B. Shabbat 67b.
30. *Mishnah,* Pirqei Avot 4:1.

A Greek Orthodox Perspective

Emmanuel Clapsis*

Philosopher Hans Jonas correctly observes:

> As long as [a] danger is unknown, we do not know what to pre-
> serve and why. Knowledge . . . comes . . . from the perception of
> what to avoid. . . . Because this is the way we are made: the per-
> ception of the *malum* [evil] is infinitely easier to us than the per-
> ception of the *bonum* [good].[1]

Certainly this has been historically true regarding the emergence of
concern for the environment. The Christian churches in the third
assembly of the World Council of Churches (New Delhi, 1961) felt
comfortable with the declaration that "the Christian should welcome
scientific discoveries as new steps in man's *dominion* over
nature"(emphasis in original).[2] It was only in the early seventies as the
result of urgent warnings from the scientific world that some churches
began to reflect on the doctrine of creation in hopes of contributing to
the resolution of the global ecological crisis. The heightening Orthodox
interest in this matter is marked by the "Encyclical on the Environment" issued by the ecumenical patriarchate in 1989.[3] The aim of this
document was to raise the awareness of all Orthodox Christians concerning the ecological crisis and their attendant Christian responsibility.
It also instituted a commemoration to be kept on the first day of September each year dedicated to the protection of the environment; local

*Emmanuel Clapsis is Associate Professor of Dogmatics at Holy Cross Greek Ortho-
dox School of Theology in Brookline, Massachusetts.

Orthodox churches were asked to offer prayers for the protection of the environment, to develop educational programs, and to exhort the leaders of the world to take all the measures for the preservation of the natural world. Since the issuing of this document, the ecumenical patriarchate has supported and actively participated in the ecological movement. Only recently Ecumenical Patriarch Bartholomew sponsored an interreligious and interdisciplinary symposium titled "Revelation and the Environment: A.D. 95–1995" in Patmos, Greece, marking the 1900th anniversary of St. John's composition of the Apocalypse, the final book of the Christian Bible. A notable outcome of this conference was the expression by representatives of different religions and Christian denominations of the need to enlarge the understanding of sin so it might be considered not merely in anthropological and social terms but also in terms of the natural world. The conference participants also suggested seven steps for action, which are given at the end of the essay.

What is the cause of this unprecedented activism among Orthodox Christians on matters of ecology? It is the awareness that human beings are destroying the natural environment to such a degree that life itself is seriously threatened. Scientists have warned that pollution of the environment and destruction of biodiversity within the natural world endanger life on earth. The excessive consumption of natural resources denies the sanctity of creation and constitutes a "rape" of nature by human beings. As Ecumenical Patriarch Bartholomew stated: "It has become painfully apparent that humanity, both individually and collectively, no longer perceives the natural order as a sign and a sacrament of God but rather as an object of exploitation."[4] Mistreatment of nature is a symptom of a spiritual illness, an anthropological heresy, since "when we become sensitive to God's world around us, we grow more conscious also of God's world within us. Beginning to see nature as a work of God, we begin to see our own place as human beings within nature."[5]

The urgency and seriousness of the ecological crisis compel churches to act. They must offer a constructive contribution to the solution of this problem, especially in view of the fact that Christianity has been accused of contributing significantly to this crisis.[6] The magnitude and urgency of the ecological crisis require that religious communities, in conversation with scientists, economists, social thinkers, and politicians, articulate ethical principles that may halt or at least limit the destruction of natural life and resources. Furthermore, as the Orthodox contribution to the Justice, Peace, and Integrity of Creation (JPIC) conciliar process points out, we must find ways to unite our efforts with all people of goodwill against all forms of injustice and disintegration that threaten the gift of life.[7] It is, however, a salutary event that the perils faced by the earth have moved theologians, scientists, economists, and politicians out of their professional and social "nests," destroying the attitude of isolationism that has existed in the past, as they together struggle to find ways to protect the environment from all destructive forces and cultural trends.

While strong denunciations of excessive consumption and use of world's resources appear in all Orthodox statements about the environment, population growth and its impact on the environment are not explicitly addressed. However, the empirical facts cannot be ignored. The population increases by three people each second, ten thousand each hour, 98 million each year. Humankind is increasing by 1 billion every decade. The present world population is approximately 5.7 billion. Are there any limits to population growth? How can one know when the world is overpopulated? If we define overpopulation as that state when the production of food can no longer keep pace with the growth in population, we have grounds to be anxious about the future, for food production has begun to decline. If, on the other hand, we consider a nation to be overpopulated when it destroys its basic natural resources, then we must consider the rich nations of the north to be overpopulated. The people of the rich nations, which represent 20 percent of the world's population, are consuming 80 percent of total resources each year. If the other 80 percent of humankind were to claim the same attitudes, lifestyle, and technology for themselves and someday achieve a level of development comparable with that of the nations of the north, the resulting ecological dangers would be immeasurable. A fair distribution of the world's resources is inextricably linked to an effective policy addressing the population problem. More than 800 million people are malnourished because of an unjust or inequitable distribution of resources that does not grant them opportunities to have the necessary means to fulfill their minimum human needs.

In an effort to limit the growth of the world's population, most Orthodox theologians would encourage the use of "artificial means of contraception" within marriage as an effective birth control device. Neither abortion nor compulsory sterilization can be accepted as a means of birth control.[8] Society has the right and the duty to inform people through education and persuasion of the importance of small families and to remind potential parents of the implications of bringing children into today's world. However, under no circumstances must the freedom of each couple to decide the size of their family be surrendered to any governmental technocrats, however benevolent they may be.

The depth of the ecological crisis is such that an ethical approach alone will not solve the problem. The ecological crisis is a cultural failure. The solution of the ecological problem is not simply a matter of management and technicalities, important as these may be. It is a matter of changing our spiritual attitudes, indeed of changing our very worldview. For it is a certain worldview that has created and continues to sustain the ecological crisis. This of course becomes abundantly clear from the historians of consumption, who have shown that the "great transformation" of the west included not just an "industrial revolution" but also a "consumer revolution."[9] The consumer revolution represents not just a change in tastes, preferences, and buying habits but also a fundamental cultural shift. It changed the Western concepts of time, space, society, the individual, the family, and the state. The retailing analyst Victor Lebow notes that "our enormously productive economy . . . demands that we make consumption our way of life, that we convert the buying and use of goods into rituals, that we seek our spiritual satisfaction, our ego satisfaction, in consumption . . . we need things consumed, burned up, worn out, replaced, and discarded at ever increasing rates."[10] Limiting the consumer lifestyle to those who have already attained it is not politically possible, morally defensible, or ecologically sufficient. And extending that lifestyle to all would simply hasten the ruin of the biosphere. If the life-supporting ecosystems of the planet are to survive into future generations, the consumer society will have to dramatically curtail its use of resources—partly by shifting to high-quality, low-input durable goods and partly by seeking fulfillment through leisure, human relationships, and other nonmaterial avenues. Scientific advances, better laws, restructured industries, new treaties, environmental taxes, grassroots campaigns can all help us reach that point. But ultimately, sustaining the environment that sustains humanity will require that we change our values, our lifestyle, and our understanding of what brings authentic human fulfillment and happiness. This means that we must radically change the prevailing understanding of the natural world and the place of human beings in it, as well as changing the prevailing notion of the relations between history and the natural world.

In a radical reorientation of our culture, the Orthodox church will reread her doctrinal teachings and liturgical ethos in an attempt to communicate her contribution to this endeavor. For Orthodox theology, the world as created reality has a beginning; it was created "out of nothing," *ex nihilo*. As such, the world is a finite reality with a beginning and an end. One could say, therefore, that the nothingness out of which the world came into being permeates it and affects every single being within the universe. "The world as a whole, like every part of it, exists under the threat of nothingness, because it was created out of nothing in the absolute sense of the world. The world possesses no natural power in itself which would enable it to overcome this situation, for if it did it would have been immortal and eternal by nature." Thus, the world's life depends upon its potential to be united with a life-giving reality other than itself. For Orthodox theology, the life of the world is derived from its unity with God, a unity that does not abolish the natural "otherness" between God and the world. The vocation of human beings, who intrinsically relate with God and the material world, is to make this unity an active reality. Because human beings form an organic part of the material world, being the highest point of its evolution, they are able to carry with them the whole creation in its transcendence through the freedom that God has bestowed upon them. It is precisely this that makes human beings responsible—in a sense the only beings who are responsible—for the fate of creation.

If the life of the world is somehow dependent on how human beings relate to it, it is possible that human beings can use the world for their own benefit not as a relational subject but as an object of gratification, a utility, or an instrument subjected to their dominion. This utilitarian attitude to creation is a logical outcome of the belief that human beings differ from the rest of creation by way of their capacity to dissociate themselves from it rather than associate themselves with it. Nature becomes the possession of human beings, a source for deriving self-satisfaction and pleasure. Science and technology represent the employment of humankind's intellectual superiority for the purpose of discovering ways and means by which human beings might derive the greatest possible profit from creation for their own purposes. Theology becomes co-responsible with science and technology for the ecological problem by its persistence in identifying the *imago dei* with human rationality.

If, however, we identify the *imago dei* with freedom, it is possible to conceive a qualitatively different relationship between human beings and nature, one that is not necessarily utilitarian, that does not reduce the world to an object that exists only for the gratification of human desires.[11] Human beings, because of the inherent gift of freedom that God has bestowed on them, have the capacity to transcend their nature and become agents of communion with realities other than themselves. In a personal relationship, human beings conceive the world as an integrated whole, a subject constitutive of their being, and their particular identity is formed not in juxtaposition to nature but in association with it. This relationality elevates the material world to the level of humankind's existence. Nature itself, in Patristic tradition, is understood not as a thing but a subject, *macro-anthropos* (an enlarged human being), while each human being is considered to be a *microcosm*. Since every human being contains all elements of reality (created and uncreated, intelligible and sensible, heaven and earth, paradise and the world, male and female) and because of the "ek-static" dimension of personhood, it is each person's human vocation to draw all creation into unity through love, to become the summit of all that exists and unite it to God through Jesus Christ. By taking the world into their hands and creatively integrating it and by referring to God, human beings liberate creation from its limitations and let it truly exist in its fullest potential for being. Of course, human beings in a personal relationship with the natural world will continue to draw from it what is required for their existence, but at the same time they would respect the integrity of the world by drawing from it not what they desire but only that which is required for their existence. Here the ascetic spirit of Christian monasticism is relevant as a signifier of a particular approach to life that presupposes that one transcends his or her own selfish will to dominate the world and use it for his or her own satisfaction. Christian asceticism aims at the subjection of individual, biological desires to the absolute primacy of personal relationship and communion.[12]

This mode of relational existence is manifested for us in the Eucharist, where the faithful offer the world to God, recognizing that he is the source of life and the "owner" of the world. In this act, creation acquires a sacredness that is not inherent in its nature but is imputed in and through the personhood of human beings. This distinguishes the Christian attitude from all forms of paganism and attaches to the human being an awesome responsibility for the survival of God's creation.

The communion of the world with God is the vocation of the human beings. The ecological implication of this is very plain: Human beings as mediators between God and the world can transfigure the environment through their acts—or they can do the exact opposite, rendering it polluted and poisonous.

The Patmos Proposals: Seven Steps for Action

In response to the Ecumenical Patriarch's invitation to participate in the symposium "Revelation and the Environment," delegates produced the following proposals for him to consider in formulating his response to the growing concern for the future of the world's environment. The delegates believe that in his support for the symposium, and the very positive atmosphere it achieved, the patriarch can act as a leader not just in his own but also in other religions to encourage global environmental awareness crucial to achieving a sustainable environment on earth.

1. *Establishment of a new sense of sin.* All religions affirm as an imperative the need to *care* for the earth and for the whole of nature. *To pollute the environment or not to take care of it should be seen as sin.* This new sense of sin extends beyond what has been traditionally considered wrong.

   This new category of sin should include activities that lead to species extinction; reduction in genetic diversity; pollution of the hydrosphere, lithosphere, or atmosphere; eutrophication of the hydrosphere, lithosphere, or atmosphere; habitat destruction; disruption of sustainable lifestyles. Specific examples discussed at the conference include the imminent extinction of the monk seals and the continued abuse of the Black Sea.

2. *Recognition and support of the rights of traditional communities.* All religions recognize that indigenous peoples are the architects and stewards of sustainable management, the guardians, and in the case of crop plants and animals, the creators of biodiversity; we urge churches, scientists, and environmentalists to *support the cause of indigenous peoples and traditional communities throughout the world.* Examples of how to do that include:

   • protection of their lands
   • protection of their ways of life
   • recognition of their property rights.

3. *Recognition of the lack of environmental knowledge in many levels of society.* The church should *encourage the development and implementation of education programs* for audiences from all schools (including Sunday schools) to adult (including seminary). Examples include:

   • creation of a "green Bible"
   • development of a series of publications linking the religions of the world to their natural heritage
   • at baptisms and weddings, stressing parental environmental responsibilities.

4. *Recognition of the crucial importance of improving the information exchange between church, nongovernmental agencies (NGOs), and government on environmental matters.* The church should *encourage efforts at planning, collaboration, and cooperation* whenever and wherever possible. Examples include:

- appointment of official environmental contacts at all levels of each church's organization, especially local churches
- support of at least one major cooperative project for the millennium, such as the rehabilitation of the Black Sea, the Aegean Sea, etc.
- encouragement of governments to involve the public in all aspects of natural resource management
- making the island of Patmos (not just its ecological center) an environmental showpiece.

5. *Responding to the need for clear leadership.* The church should take positive steps to establish sustainable and environmentally friendly land-use practices, resource use, and investments. Examples include:

- issuing guidelines for sustainable use and best practice for all church lands
- making sure all new or renovated religious edifices utilize the most efficient energy conservation and waste reduction and disposal technologies
- establishing nature reserves or afforestation projects on all appropriate church properties
- putting all church investments should be in "green" or environmentally friendly sustainable businesses or investment portfolios.

6. *Recognition of the vital role that the world's media could play in promoting awareness of environmental issues.* The church should *encourage the media to feature environmental issues on a more sustained and regular basis.* Examples include make space on a regular basis to encourage environmental awareness and sustainable activities in all publications, from parish newsletters to national and international press and television.

7. *Recognition of the urgency of the earth's environmental problems.* Projects promoting the Patmos proposals are of utmost urgency. They could include:

- establishment of a Green Award(s) by the church for, e.g., a scientist or environmental activist
- development of a church outreach environmental agency (an Orthodox NGO)
- institutionalization of a celebratory tree planting (for births, weddings, deaths).

- encouragement of all faiths to promote a green will or trust option
- establishment of a cruising environmental educational vessel that would promote the Patmos proposals.

Notes

1. Hans Jonas, *The Imperative of Responsibility* (Chicago: University of Chicago Press, 1984), p. 27.
2. A helpful summary of the history of ecumenical explorations of creation is offered by Lukas Vischer, "The Theme of Humanity and Creation in the Ecumenical Movement," in *Sustainable Growth—A Contradiction in Terms? Economy, Ecology, and Ethics After the Earth Summit* (Geneva: Visser't Hooft Endowment Fund for Leadership Development, available through the WCC, 1993).
3. Ecumenical Patriarchate, *Orthodoxy and the Ecological Crisis* (Gland, Switzerland: World Wide Fund for Nature International, 1990); for theological reflections on ecology, see: John Zizioulas, "Preserving God's Creation: Three Lectures on Theology and Ecology," *King's Theological Review* 12 (1989), pp. 1–5, 41–45, and 13 (1990), pp. 1–5; Gennadios Limouris, ed., *Justice, Peace and the Integrity of Creation—Insights from Orthodoxy* (Geneva: WCC, 1990); Elias Economou, *An Orthodox View of the Ecological Crisis,* (Athens 1990); Elias Economou, *Theologike Oikologia-Theoria and Praxis* (Athens, 1994) in Greek; A. Keselopoulos, *Man and the Natural Environment* (Athens, 1989) in Greek; Vasileios Giultsis, "Creation and the Ecological Problem," in Gennadios Limouris, ed., *Come, Holy Spirit, Renew the Whole Creation,* (Brookline: Holy Cross Orthodox Press, 1990), pp. 228–233; Paulos Gregorios, *The Human Presence: An Orthodox View of Nature* (Geneva: World Council of Churches, 1978); Paulos Gregorios, *Cosmic Man: The Divine Presence—The Theology of St. Gregory of Nyssa* (New York: Paragon Press, 1988); Paulos Gregorios, *Science for Sane Societies* (New York: The Paragon Press, 1987); Philip Sherrard, *The Eclipse of Man and Nature: An Enquiry into the Origins and Consequences of Modern Science* (West Stockbridge, Mass.: Lindisfarne Press, 1987).
4. Patriarchal "Encyclical on the Environment," September 1, 1994.
5. Ibid.
6. For a critical assessment of the ecological critique of Christian theology, see H. Paul Santmire, *The Travail of Nature: The Ambiguous Ecological Promise of Christian Theology* (Minneapolis: Fortress Press, 1985), pp. 1–12.
7. Gennadios Limouris, ed., *Justice, Peace, and the Integrity of Creation-Insights from Orthodoxy* (Geneva: WCC, 1990), p. 6.
8. Stanley Samuel Harakas, *Living the Faith, the Praxis of Eastern Orthodox Ethics* (Minneapolis, Minn.: Light and Life Publishing, 1992), p. 133.

9. Fernad Braudel, *Capitalism and Material Life 1400–1800* (London: Weindenfield and Nicolson, 1973); Neil McKendrick, John Brewer, and J. H. Plumb, *The Birth of a Consumer Society: The Commercialization of Eighteenth-Century England* (Bloomington: Indiana University Press, 1982).

10. This quote and much of my thought on consumption is indebted to Alan Thein Durning's article on consumption in *Sojourners* (August 1994), pp. 20–23. For a detailed description of his views, see his book *How Much Is Enough? The Consumer Society and the Future of the Earth* (New York: W.W. Norton, 1992).

11. John Zizioulas, "Preserving God's Creation: Three Lectures on Theology and Ecology," pp. 1–5, 41–45: 13 (1990), pp. 1–5.

12. Christos Yannaras, *The Freedom of Morality* (New York: St. Vladimir's Seminary Press, 1984), p. 218.

An Islamic Perspective

Abdul Cader Asmal and Mohammed Asmal*

To appreciate the context within which Muslims view the topic under consideration, it may be helpful to understand that for Muslims everything emanates from God. God is Supreme and One, is Eternal and Infinite, is Omnipresent and Omniscient, is Benevolent and Merciful, and is the Creator and Sustainer. God is the Creator of our universe, of our sun that sustains us, of our planet, and everything on it—the oceans, the terrain, the vegetation, and the animals including humans.

As the Creator, God has a plan for all of His creation and creatures, which in turn are responsive to His laws, as the following verses from the Qur'an testify:

- "Certainly, to Him belongs all that the heavens and the earth contain" (22, 16).
- "Do you not see that it is God who has created the heavens and the earth, not without a plan and a purpose" (14, 19)?
- "All that is in the heavens and the earth praise Allah, the Sovereign, the Holy, the Mighty, the Wise" (62, 1).
- "Do you not see that all things bow down to Allah, all things in the heavens and the earth, the sun, the moon, the stars, the hills, the animals; and a great number of mankind" (22, 18)?

*Abdul Cader Asmal is chairman of communications for the Islamic Council of New England. Mohammed Asmal is a 1995 graduate of Harvard College, a former officer of the Harvard Islamic Society, and editor of the Harvard *Crimson*. Asmal is currently involved in HIV research and is working toward a joint M.D./Ph.D. degree at Columbia University. This paper was originally given at the AAAS/BTI conference, which is the basis for this volume.

In His scheme of Creation, God has distinguished humanity from the rest of His creatures by two attributes:

1. He has endowed humankind with a free will, together with the intelligence to reason, and a guidance to recognize good from evil.
2. He has made humankind His trustee on earth with the responsibility to respect and preserve God's creation according to His laws. In the words of the Qur'an, "It is He [Allah] who has made you his vice-regent on earth" (6, 165).

Because Islam is a comprehensive way of life, with God-consciousness permeating all of human activity, whether it be business, education, social conduct, or science, religion and science become inextricably bound in God's creation. Thus, from an Islamic perspective, the environment, consumption, and population are no different whether viewed from a religious or a scientific angle.

The ultimate objective of God's law is the common good of His entire creation, as illustrated by the following Qur'anic verse: "We created man and gave him the faculty of speech. The sun and the moon rotate in ordered orbits, the plants and the trees, too, do obeisance. The firmament—He raised it high, and set the balance of everything, so that you [humanity] may not upset the balance. Keep the balance with equity, and fall not short in it" (55, 3–9).

These powerful Qur'anic injunctions to humankind to preserve the environment are amplified by the exhortations of Prophet Mohammed in the following Hadith[1]:

- "Whoever brings dead land to life, for him is a reward in it, and whatever any creature seeking food eats of it shall be reckoned as charity from him" (Mishkat al-Masabih).
- "The world is green and beautiful and God has appointed you as His stewards over it. He sees how you acquit yourselves" (Muslim).
- "Whoever plants a tree and looks after it with care, until it matures and becomes productive, will be rewarded in the hereafter" (Bukhari and Muslim).

Abdullah Omar Naseef, secretary general of the Muslim World League and a staunch advocate of environmental protection, observes that "ecological effects know no boundaries, are no respectors of national frontiers," and elaborates on the implications of being God's vice-regent on earth: "The terms of this trust mean that mankind has no absolute right to use, abuse or dispose of the environment and God's creation according to only the needs of today, or human whims. The essence of our stewardship is accountability for the continued well-being of the earth itself and all the species of flora and fauna that live thereon."

Turning now to the subject of consumption, the Qur'an is again extremely explicit with respect to "keeping the balance with equity" (55, 3–9) and reminds humankind of this obligation in many verses:

- "God bestowed His blessings on the earth, and measured therein sustenance in due proportion . . . in accordance with the need of those who seek" (41, 10).
- "Verily! your Sustainer augments or circumscribes the means of sustenance for whomsoever He pleases—for He knows and keeps a watchful eye on all His creatures" (17, 30).
- "Eat and drink, but waste not by indulging in excesses. Surely, God does not approve of the intemperate" (7, 31).

The Holy Prophet has himself set the standard by which the usage of anything should be judged: "The merit of utilization lies in the benefit it yields, in proportion to its harm" (Hadith recorded by Tirmidi).

Naseef has once again very eloquently summarized the delicate relationship between natural resources and their utilization when he stated, "We need to define a proper balance between meeting human needs and sustaining the natural world. We must seek an appropriate answer to the conundrum of harmonizing the exploitation of the riches of the natural world for human use with husbanding these resources for the use and enjoyment of future generations."

Our assault on the environment with the consequences we now face is somewhat more dramatically expressed by Fazlun Khalid, another Muslim ecologist: "Our self-indulgence has led us to compete with each other as consumers, as individuals and as countries sucking things out of the earth at an ever-increasing rate and discharging a level of waste which the earth cannot recycle, thus contributing to the rapid destruction of the habitats and lifestyles of the weakest amongst us. We are rampaging through the delicate balance of nature. Savaging other species to extinction. Robbing future generations of their inheritance. We have become so trapped in our own self-indulgence we are not even aware of it."

Turning now to the third and vital issue of population, which within the context of our presentation I infer to be a reference to "population control," the Qur'an has this stricture: "Kill not your children for fear of want; We shall provide sustenance for them as well as you. Verily, killing them is a great sin" (17, 31). These verses refer specifically to the abominable practice of female infanticide that was practiced in Arabia at the time of the revelation. Although the Qur'an makes no categorical statement either in favor of or against contraception, most Muslim scholars over the years have promoted the view that Islam discourages if not forbids contraception, basing their opinions on the extrapolation of the verses that condemn infanticide.

If the Qur'an has not directly addressed the issue of contraception, the Prophet himself debated the question in a variety of contexts with his disciples. His conclusion was that contraception with the then available means was permissible, with the caveat that if God so intended, no form of contraception could thwart His Will.

On the related issue of abortion, Islam has a clear and unequivocal position: Under no circumstances can abortion be practiced as a method of population control. On the other hand, it is permitted in Islam whenever medically indicated, and always if the life of the mother is endangered.

In summary, while neither the Qur'an nor the Hadith forbids the practice of contraception, most schools of Islamic jurisprudence seem to have interpreted it as an undesirable, or *makruh,* act (Ebrahim 1989). Whether this is indeed the case is not as important as the reaction such a belief engenders. Thus, when the government of Pakistan in the late 1960s attempted to introduce a national family planning policy it met with stiff opposition from the Ulama (religious scholar), whose objections to the idea are listed as follows (Ebrahim 1989):

• Family planning is akin to infanticide.
• Family planning is unnatural or contrary to human nature.
• Family planning is based on disbelief in the providence of Allah.
• Family planning would be tantamount to ignoring the prophet's wish that Muslims should increase in number.
• Family planning will lead to disastrous social consequences.
• Family planning is a sort of conspiracy of the Western imperialists against the developing nations.

This single example of resistance to family planning highlights the stark contrast between the vision with which traditional Muslim scholars and the newer proponents of an "Islamic science" have focused on the environment and its protection and the virtual myopia they display when it comes to tackling the pressing if contentious issue of population control. The apparent paradox will be taken up later in this essay.

The challenges that Muslim scholars face in promoting what they regard as God's manifesto for His creation arise from both within and without, are general and specific, and can be discussed under four categories:

1. implementation of the "known or accepted"
2. reinterpretation (*Ijtihad*) of the "less known or accepted"
3. legitimization of an "Islamic science"
4. dismantling of the "Islamic" stereotype

Perhaps the most dispiriting contradiction that Muslim scholars have to face is clearly defined and comes from within—and that is the reluctance or the impotence of the leaders of so-called Muslim countries to implement, at least within their own domains of influence, the undisputed Islamic guidance with respect to the environment and its conservation. Implementation of what is accepted calls for neither Ijtihad, nor the recognition of an Islamic science, nor the need to combat stereotypes.

Ijtihad, or the reinterpretation of the immutable laws of the Creator from their original sources to meet the changing circumstances of the present age, is a much more challenging task for the present-day scholars of Islam.

The starting point for all Islamic views of the world is the Qur'an. As the direct revelation from God to the Prophet Mohammed, the Qur'an is the primary source for Divine guidance. In addition, the Prophet established a model for behavior through his everyday actions and speech. His words and deeds were recorded in several corpuses known as the *Hadith*. Together the Qur'an and the Hadith form the textual sources of Islamic law, with the Qur'an taking precedence and the Hadith acting supplementarily.

The Qur'an and Hadith contain both universal and contextual elements. While there is certainly a strongly timeless message within the Qur'an, parts of the message are addressed specifically to Mohammed and his followers of that day and age. The challenge for the Islamic legal scholar and moralist was and is to distinguish the contextual elements in the Qur'an from the broader universal message. Moreover, because Qur'anic verses may be allegorical and have many layers of meaning, interpreters have to guard against projecting both cultural subjectivity and individual creativity onto their understanding of the text. In addition to the primary texts, scholars rely on the use of analogy and consensus taking to derive Islamic law.

Islamic laws thus derived are not rigid but flexible. With respect to this flexibility, Aziza Al-Hibri, a Muslim lawyer and activist notes, "It is an essential part of Qur'anic philosophy, because Islam was revealed for all people and for all times. Consequently, its jurisprudence must be capable of responding to widely diverse needs and problems." Among the fundamental principles of Ijtihad are the following:

- Laws change with changes in time and place.
- Choosing the lesser of two harms.
- Preserving public interest.

Application of these principles as we have seen is perfectly consonant with respect to the environment and its preservation. What about population control? Why is it that though the Qur'an is silent on the subject of contraception, and the Prophet himself allowed it, Muslim scholars have adopted the position that family planning is disapproved of if not prohibited. Two pieces of evidence are consistently cited in support of this conclusion: the Qur'anic prohibition of infanticide out of fear of want, and the Prophet's exhortation to his fledgling community to multiply.

In formulating a policy on population control, present-day scholars of Ijtihad have to be cognizant of the fundamental principles enunciated by Al-Hibri, as noted previously, within the framework of the special value attached to family life in Islam, but also against the wider backdrop of the devastation wrought in many parts of the Muslim world by unrestricted population growth causing a downward spiral of poverty, disease, starvation, illiteracy, and degradation. In order for Muslim scholars to address this issue with urgency and earnestness, they have to recognize it as a Muslim problem and not one created by some external force bent with evil machinations to subvert Islamic value systems. If these scholars truly believe in the viability of an "Islamic science," they should be able to isolate the technological advances of family planning as suited to a Muslim family and discard the seductive trappings antithetical to the Islamic concept of family life that may be promoted with the package, such as extramarital promiscuity and abortion on demand. If Islamic science is indeed going to be promoted as a value-added science, as Naseef suggests when he says, "We need to learn to use the value system of Islam to distinguish between the beneficial and harmful effects of science and technology," then, though these remarks refer to the environment, their implications would have an immeasurably greater relevance when extended to the subject of population control. The acceptance of such a proposition poses the supreme challenge to the proponents of an Islamic science.

However, as they evolve their neophyte paradigm of an Islamic science, Muslim scientists face an onslaught that could obliterate their efforts even before the prototype is unveiled. The source of this onslaught is epitomized in the words of Zdenek Kopal, a German astronomer, in an essay in *Islamic Cultural Identity and Scientific-Technological Development*: "A Western student of science is inclined to regard the Holy Scripts of any religion only as a mirror of the times at which they were conceived. For him the only Holy Writ is the open book of nature. Modern students of nature are highly skeptical of revelation as a source of wisdom or method of inquiry; and apt to regard any such ideas as an illegitimate shortcut. Science is indeed that part of our culture which sets its goals in the future, and does not search for the answers to its problems in past revelations."

Such views are diametrically opposed to those espoused by proponents of an Islamic science. As Naseef puts it, "The dominant way of understanding science and technology poses a challenge to Islam. The Muslim intellectual must answer the question of how do we set about defining a conception of science and technology which is not secular, materialistic, reductionist and destructive of the humility of mankind before its Creator." Seyyed Nasr, a renowned author and ardent proponent of an Islamic science, elaborates further: "So extensive is the dislocation caused by the spread of the teaching of a science of nature without reference to God (vis-à-vis the ecological crisis and the danger of nuclear war) that one cannot, in all honesty, simply claim that there is no problem by claiming modern science to be the knowledge which Islam encourages or by ignoring modern science as if it did not exist. Modern science must be taught and known well for what it is, without raising it to an idol of the mind or forgetting its essentially agnostic and secularist nature. Islamic science, while incorporating all the factual discoveries of modern science, would relate these facts to higher principles and would remain aware of the ultimate cause of all things which is God."

If the legitimization of an Islamic science is a specific challenge to some Muslim academicians (and it is by no means certain that to most Muslim intellectuals this is indeed a prerequisite vehicle for the delivery of an Islamic perspective), that of dismantling the negative stereotypes is close to the heart of all Muslims. Muslims would like the rest of the world to understand that for them Islam is not just a religion but a comprehensive way of life, of which science is but one facet. All of God's creation is subject to the same laws, science being no exception. For Muslims to get out this all-encompassing message of Islam, they first have to practice what they believe in, and by their behavior become exemplars to the rest of the world. It is only through this that they will be able to narrow the chasm between the ideals of Islam and the shortfall of Muslims. In concert with their own self-improvement, Muslims have to dismantle the stereotypes within which they have become standardized.

In this context, it is ironic that the world feels more threatened by a nonexistent "Islamic bomb"[2] than by the reality of a superabundance of existing warheads, not exactly under the control of the most stable of world leaders, that could annihilate us many times over. From a truly Islamic perspective, not only is such a weapon of mass indiscriminate destruction total anathema to Islam's view of creation, the environment, its creatures, and among humans the noncombatants, but it is also a luxury that no Muslim country can afford and a goal these countries do not need to create. Sadly, both India and Pakistan (the former with a significant Muslim minority and the latter a so-called Islamic country) have indulged in this godless venture.

For Muslims to be able to transmit the true and positive messages of Islam to the world, they need to dispel the negative images that overshadow anything they now propose. When they do, it will be evident that Senator Al Gore's admonition of an "earth in the balance" is but the echo of an injunction expressed in the Qur'an some fourteen centuries ago: "We . . . set the balance of everything. Keep the balance with equity and fall not short in it." However, for the rest of humanity to share in Nasr's vision "that the thinkers of the Islamic world will be able to transform the modern sciences into sciences which, while providing knowledge of the created order, will also remind man constantly that 'Wherever ye turn there is the Face of God'" (2, 115), Muslims will have to do more than just break down stereotypes. At the very least, Muslims will have to emancipate themselves from their selective moratorium on the Ijtihad of key issues. For instance, Muslims must view population control from an Islamic perspective by taking into account its other imperative: social equity and justice.

In the meantime, can we envision any equity? The following maxim, whether derived from the sayings of Mohammed, Jesus, or Buddha, inscribed as a humble bumper sticker, exemplifies all their lives and does offer us a vision, "Live simply, so that others may simply live."

Notes

1. Hadith are statements of the Prophet Mohammed authenticated and recorded by scholars of unquestionable integrity. These Hadith are included after the following quotations.
2. With India and Pakistan joining the nuclear club, an "Islamic" and "Hindu" bomb join the family of "Christian" and "Jewish" ones.

References

Al-Hibri, Azizah. "Family Planning and Islamic Jurisprudence." In *Religious and Ethical Perspectives on Population Issues*. Washington, D.C.: Religious Consultation on Population, Reproductive Health and Ethics, 1993.

Ebrahim, Abul Fadl Mohsin. *Abortion, Birth Control and Surrogate Parenting: An Islamic Perspective.* Indianapolis: American Trust Publications, 1989. This book focuses on the teachings of the Qur'an and Hadith as they pertain to the beginnings of human life and the sanctions for the termination of human life.

Gottstein, Klaus, ed. *Islamic Cultural Identity and Scientific-Technological Development.* Baden-Baden: Nomos Verlagsgesellschaft, 1986. In particular, see the article by Zdenek Kopal.

Khalid, Fazlun M., and Joanne O'Brien, eds. *Islam and Ecology.* London: Cassell Publishers, 1992. This book addresses issues as wide ranging as animal husbandry, desert reclamation, international trade, and science from a Muslim perspective.

Sardar, Ziauddin, ed. *An Early Crescent: The Future of Knowledge and the Environment in Islam.* London: Mansell, 1989. This book brings together a number of Muslim intellectuals, who speculate on the future of knowledge, including that of the environment from a Muslim perspective. Articles by Abdullah Omar Naseef and Seyyed Hossein Nasr are cited above.

Toward the Revival and Reform of the Subversive Virtue: Frugality

James A. Nash*

One of the most interesting and largely unexamined problems in modern Christian ethics is the major demotion of frugality as a personal and social norm. An old and honored virtue that was once near the heart of Christian economic ethics has become one of the most neglected—and unnerving—norms in modern morality. Though popular appeals for frugality are uncommon but possibly increasing,[1] ethically substantial interpretations and justifications of frugality in Christian contexts seem quite rare.[2] The decline of frugality is, however, much more than a matter of intellectual curiosity; it is a serious moral problem in itself if frugality is socially and ecologically significant.

Historically, frugality was a prime Christian economic norm. It was, for example, a prominent practice in the patristic and monastic traditions of East and West—though voluntary poverty sometimes went beyond the moderation characteristic of frugality to various degrees of austerity. Moreover, frugality combined with industry, honesty, reliability, equity, generosity, piety, and covenantal solidarity constituted the core of the classical "Protestant ethic," which Max Weber described, with some exaggeration, as "worldly asceticism."[3] Frugality characterized not only Calvinism but virtually all of the Protestant traditions—Baptist, Anabaptist, Quaker, Anglican, Wesleyan, Lutheran, etc., despite some significant differences in economic ethics. These Catholic,

*James A. Nash is former executive director and is now senior scholar of the Churches' Center for Theology and Public Policy in Washington, D.C.

Orthodox, and Protestant perspectives reflect the fears about the spiritual perils of prosperity and the commitment to frugal consumption "in the service of love" evident in the churches of the New Testament.[4] In gratitude to God, frugality was a rejection of vanity and envy and an expression of fidelity and relationality. It was part of godliness or sanctification.

The norm of frugality—often in the guise of "simplicity"—exercised a strong but ambiguous and fluctuating influence on U.S. history. It was a prominent component in "a sustaining myth of national purpose," but it existed alongside a counter myth of progressive prosperity. Frugality was often honored more as an ideal than as a practice.[5]

Yet in contemporary cultures that celebrate the prospects of progressive plenty, even the ideal is banished. Frugality is often greeted with amusement, ridicule, or even contempt. It is portrayed as unfashionable, unpalatable, and even unpatriotic. It is considered the essence of an antiquated and much berated Protestant ethic, symbolized by the painting *American Gothic*. Frugality is the "miserly virtue," according to Lord Keynes, appropriate only to historical conditions of scarcity and now anachronistic under conditions of high productivity.[6]

In Christian contexts, frugality has certainly not been fully forgotten, but it has been significantly demoted—probably reflecting an accommodation to cultural values or an adaptation for the sake of prudent prophecy. Frugality is not featured, for example, in most modern manuals in Christian economic ethics. It and its closely allied concepts (except temperance, and then not in terms of economic production and consumption) are not listed and are barely discussed in other entries in the popular *Westminster Dictionary of Christian Ethics*[7] or even *The Encyclopedia of the Reformed Faith*.[8] Concerns about overconsumption are not prominent in the U.S. Catholic bishops' economics pastoral (though they are present in two paragraphs, one under the "call to holiness" addressed specifically to Christians).[9] Essentially the same is true of various Protestant economic statements. A study paper of the United Church of Christ, for example, discusses the problem of scarcity and hints at frugality in complaints about consumerism and appeals to Christians for personal stewardship.[10] Frugality, under the name of "sufficiency," however, is a serious theme in a recent study document on ecological responsibility from the Presbyterian Church (USA).[11] Thus, frugality remains as an undercurrent in contemporary Christian ethical and ecclesiastical thought, but that undercurrent generally flows with dramatically less power than it did historically.

In evaluating this moral shift, it must be noted at least that a claim to continuity with various traditions of Christian economic ethics seems inconsistent with the abandonment or neglect of one of the prominent norms of these traditions, *unless* one can argue that said norm is a time- and culture-bound standard that is no longer relevant in an age of plenty. Such arguments, to my knowledge, have not been forthcoming, and such arguments, in my view of the current economics–ecology dilemma, would not seem defensible.

This chapter offers an ethical interpretation and justification of frugality from a Christian perspective. I argue that frugality is not an anachronistic or innocuous norm—an "old-fashioned" or "garden variety" virtue. On the contrary, frugality is a richly relevant and potentially transformative standard to combat excessive and unfair consumption *and* production. Solutions to major social and ecological problems depend on the revival of this virtue and its re-formation from a personal virtue into a social norm—as, indeed, it was intended to be in Puritanism[12] and its predecessors. The frugality I interpret and defend is an ethically conscious form, intentionally responsive to social and ecological conditions, rather than an unexamined "way of life" of individuals or communities. Not all forms of frugality can be described as ethical choices.

Frugality as Economic Subversion

I must first offer an interpretation of the contemporary economic and ecological context to which an ethically conscious frugality is a dissent or protest. A modern effort to revive this norm makes adequate sense only in this context.

Against this background, frugality is the subversive virtue, because it is a revolt against an economic system that depends upon intensive production and consumption to keep the system going and growing. Frugality is an encounter with an economics ethos that cannot afford frugality if that ethos is to thrive, and it is an effort to resist, even undermine, the central assumptions of that ethos. Four basic and interrelated characteristics of this "revolt" or subversion are important to note here.

First, frugality rejects the popular assumption that humans are insatiable creatures, ceaselessly acquisitive for economic gains and goods and egoistically committed to pleasure maximization.[13] Under this assumption, frugality as a virtue is irrational, even unnatural. The function of economic enterprises is to respond to the indefinitely elastic demands of these mythical humans—and to re-create real humans in that image by constantly stimulating demands.

Yet, the practice of frugality refutes this simplistic moral anthropology and its self-fulfilling prophecy. It bears witness to the fact that humans are not reducible to instinctive bundles of insatiable appetites and that excessive and unfair consumption is not so much a reflection of nature as a creation of culture. We *do* have the moral capacities to control and distribute consumption. Frugality reveals another side of human potential, as relational beings with powers of responsible control.

Second, frugality resists the temptations of consumer promotionalism—particularly the ubiquitous advertising that pressures us through sophisticated techniques to want more, bigger, better, faster, newer, more attractive, or "state of the art." We are bombarded with an abundance of commercials creating dissatisfactions and peddling wares that promise to enhance our excitements and powers, reduce our anxieties, and fulfill our aspirations. Commercials encourage impulse and therapeutic buying to improve self-esteem and social status. They stimulate feelings of envy and inadequacy unless we own given products. The more and bigger we can buy, the "better off" we are, since "consumerability" is the measure of meaning. Indeed, the more sales we boost, the "better off" the national and even world economies will be.

This promotional system not only caters to consumer wants; more important, it creates them, in order to provide goods and services to supply these demands. The dynamic seems to be fed more by the needs of competitive production than by the actual wants of consumers. Shopping—accelerated by instant credit—is now the main means of recreation for millions. The "pursuit of happiness" has become the quest for acquisitions. "Malling" has become part of the North American ethos—indeed, a tourist enterprise, and the malling of the planet seems to be approaching.[14] The result is not only an abundance of benefits, but also a wealth of waste, irrelevancies, and perils to human welfare.

In contrast, by resisting these temptations and deceptions, frugality is again a witness to the fact that humans are far more than manipulable consumers. It is an affirmation of human dignity—our moral potential, our deepest yearnings, our status as ends, not merely means—against the onslaught of mass manipulation. It is also an expression of purchasing power, with some potential for redirecting economic production to better means and ends.[15]

Third, frugality struggles against the various psychological and sociological dynamics, beyond market promotionalism, that stimulate overconsumption. The causative factors in overconsumption appear to be numerous, complex, and interwoven. To some degree, they are self-generated, but they are also, and probably more so, culturally shaped. Certainly greed and gluttony, envy and prodigality are implicated, but the causes are deeper and sometimes more sympathetic than these vices suggest.

Some of the main reasons for overconsumption appear to be as follows:

1. A social ideology of rising expectations, driven by market forces and the belief that the good depends on more and better goods, creates feelings of relative deprivation and persistent discontent with present possessions.[16]

2. Changing socioeconomic conditions and technological developments continually create new "needs"—electronic devices, for example—to function effectively and decently in the culture.

3. Higher discretionary income for many (despite stagnant or falling real wages for many others) has made possible extravagant expenditures.

4. Shopping has become an addiction for many, compelling fixes of novelty with short-lived and diminishing gratifications.[17]

5. Ever present hedonistic impulses are culturally encouraged, thus sanctioning and stimulating the quest for self-indulgent comforts and conveniences, new thrills and immediate pleasures.

6. Consumption patterns for many may be compensation for loneliness, powerlessness, insecurity, and ultimate vulnerability, given the decline of communal supports and transcendent values.[18]

7. Conformity to the values of one's reference group may stimulate excessive consumption in order to convey one's equal worth or status—even to the point of serious indebtedness for many. Conforming or imitative consumption is often accompanied by the fear of social ostracism or the loss of social acceptability for the unfashionable.[19]

8. Competitive consumption relative to one's reference group flaunts success and superiority, power and importance, and it often creates a vicious spiral of ostentatious waste and indulgence as the competitors seek to surpass one another. This syndrome is the "conspicuous consumption" analyzed nearly a century ago by Thorstein Veblen.

The human reasons behind overconsumption are not always or even usually morally contemptible; they are often merely mournful, revealing the hollowness and stress of many affluent lives. They reflect the quest for self-esteem, social acceptance, personal satisfaction, and ultimate meaning, while being culturally conditioned to follow trails that frustrate these hopes. In contrast, frugality finds self-esteem, satisfaction, and meaning in resisting cultural pressures, curbing desires, and reducing consumption. Frugality is more than an alternative lifestyle; it is a counter vision of being a purposeful and responsible human being in a purposeful and responsible society.

Fourth, ethically conscious frugality rejects the prevailing ideology of indiscriminate, material economic growth. At this point we confront a fundamental argument of nearly all the prominent contemporary advocates of ethically conscious constraints on production and consumption, including Herman Daly, John Cobb, William Ophuls, Christopher Lasch, Ian Barbour, Lester Milbrath, Theodore Roszak, Erazim Kóhák, Juliet Schor, Paul Wachtel, Lester Brown, and Alan Durning.[20] From these perspectives, unconstrained material economic growth is a strategy for living beyond biophysical means—to the detriment particularly of poor people and nations, future generations, and nonhuman species. While this growth model provides a variety of values—from multiple millions of jobs to vast tax revenues, from many important goods and services to philanthropic benefits—it also functions self-destructively through excessive and unfair use. It is a major factor in destroying the ecosystems on which the well-being of all social and economic systems finally depends. It presumes the practical inexhaustibility of the products and capacities of nature and human autonomy from the rest of nature. In reality, however, we live in an age of "ecological scarcity"[21]—prospectively at least, actually on many particulars. Everything is limited—from the biophysical resources necessary for human thriving to the technological powers to extend biophysical limits. Present patterns of economic activity appear to be approaching or surpassing some of these limits, locally and/or globally, in both nonrenewable resources (e.g., fossil fuels and some industrially significant minerals) and renewable resources (e.g., fisheries, forests, and croplands), as well as in wastes produced. This pattern of superproduction and superconsumption is neither biophysically sustainable nor ethically acceptable.

The emphases on biophysical limits and the economic infringements on these limits are foundational perceptions of empirical conditions in major contemporary rationales for frugality. These emphases, like antagonistic interpretations, depend not only on empirical assessments but also on value assumptions—for example, the meaning of full human development and the material conditions necessary for it. By reading reality through different value lenses than those of their opponents, socially and ecologically sensitive frugalists argue for a new economics paradigm and for frugality as a necessary condition for the ethical acceptability of that paradigm.

Thus, frugality is a revolt against some basic values of the Sumptuous Society. For the sake of personal, social, and ecological well-being, frugality rejects the gluttonous indulgence, compulsive acquisitiveness, conforming and competitive consumerism, casual wastefulness, and unconstrained material growth promoted by the peddlers of economic "progress"—and embraced to different degrees by all of us who have known the enticements of affluence.

With this interpretation of the economic and ecological context of frugality in mind, I can now offer an interpretation of frugality itself.

What Is Frugality?

What is frugality? Frugality denotes moderation, temperance, thrift, cost-effectiveness, efficient usage, and a satisfaction with material sufficiency—similar to the "contentment" celebrated in the first Pauline letter to Timothy (6:6–10). As a norm for economic activity, for both individuals and societies, frugality means ethically disciplined production and consumption for the sake of some higher ends. It is a means, an instrumental value, not an end in itself. As such, frugality thrives on conscientious conservation, restrained consumption, optimal efficiency, comprehensive recycling, and an insistence on built-in durability and repairability. Overall, frugality should probably be understood as one dimension of stewardship, but if so, that dimension should be explicit and prominent, which it is not in most popular interpretations of stewardship. Indeed, if present at all, frugality is usually fig-leafed with a generic appeal for a stewardship without specified standards and responsibilities.[22]

Generally, in contemporary thought, frugality is interpreted as strictly a personal virtue—a morally excellent trait or habit of individual behavior. A virtue issues in voluntary action. Not only means and ends but right motives matter. As a virtue, frugality is a feature of individual character ethics.

But frugality should not be so constricted. Indeed, it was not in classical thought. Both Aristotle and Aquinas, for example, interpreted the virtues as having both personal and political dimensions, since humans are not atomistic entities but relational beings.[23] Thus, both frugal persons and frugal societies can be moral realities. Yet when frugality is a social norm, emphases shift: the focus is on social ends and consequences rather than individual motives, on public decisions and structures more than personal character, on coercion as well as consent. Frugality is then a personal virtue only for those individuals who conscientiously practice it apart from social sanctions.

Nevertheless, both personal virtues and social norms are ethically necessary. In fact, voluntary commitments and social stimuli function interactively. Through incentives and disincentives, social structures can be a moral deterrent, making it difficult, if not impossible, for individuals to act virtuously; or they can be a moral catalyst, encouraging and enabling virtuous behavior. The simple act of recycling, for example, is not practical without recyclable products, recycling centers, public acceptance, and market potential, enabled by public policy. Frugal persons can flourish and function effectively only with the institutional supports of frugal societies.

Moreover, public policy in a participatory society depends on a dialectic of consent and coercion. A sufficient number of citizens must be virtuous, or willing to be virtuous, in order to have sufficient public consent to establish structures of coercion. In turn, such structures of coercion cannot only change public behavior, but, if perceived as reasonable and just, can also shape personal values and virtues, thus enhancing the level of consent. Thus, frugality, like any other operative norm, depends on both moral education and public sanctions, on both character formation and social regulation, each shaping the capacities of the other.

Some might argue that temperance is a better word for the perspective and practices I am describing. This argument has considerable historical merit. In classical philosophy and theology, temperance was one of the four cardinal "natural" virtues. In Aristotle and Aquinas, it was the disciplined self-mastery (sophrosyne) of the bodily pleasures of touch and taste.[24] Like frugality, it referred to the mean between two vices or excesses, deficiency and indulgence.[25] It was the opposite of gluttony and greed, two of the seven "deadly" or principal sins in classical thought. Unfortunately, temperance was stripped of its classical meaning by some of my Protestant ancestors in their Prohibition campaign of the early twentieth century. Popularly, the word became synonymous with abstinence from "beverage alcohol." Consequently, temperance no longer has the rich breadth of meaning that it had in classical thought—though even then temperance referred generally to moderating the desires for food, drink, and sex.[26] The last is certainly not what I would include under the category of frugality! Thus, frugality seems to be the preferable word and concept here, because, unlike temperance, the meaning of frugality is generally restricted to my concern with economic activity. Philosophically, however, perhaps frugality can be understood as the economic subspecies of temperance.

Similarly, some might argue that "sufficiency" is a better term. This argument, too, has merit.[27] Frugality certainly includes regard for sufficiency, but this word does not seem to comprehend the full economic connotations and consequences of frugality. In fact, frugality may be a less popular concept than sufficiency precisely because the former is much clearer and more realistic than the latter in conveying the disciplined constraint and even sacrifice that are necessary as an adequate response to the economics–ecology dilemma. The same criticism applies, but even more forcefully, to Theodore Roszak's use of "plenitude" to give disciplined consumption honorific connotations.[28] Nevertheless, sufficiency and plenitude are useful concepts, perhaps best understood as ends to which frugality, as an instrumental norm, is the means.

The problem with verbal substitutions in this case seems more than semantic. It is rather that a lot of the proposed substitutions for frugality suggest a dilution of moral responsibilities. The common negative reactions to the word "frugality" are often really rejections of the positive moral values the word connotes.

Distortions of Frugality

Frugality has often been the victim of "guilt by association." It has been linked to values and practices that are not inherent characteristics of frugality and that, in fact, distort its moral significance. The revival and reform of frugality depend on a dissociation from these distortions. Several of these merit some discussion to indicate what frugality is *not*— and thus to clarify what it is.

Frugality is not austerity—though if frugality is not chosen under certain conditions, biophysical limitations may impose austerity. For the balanced John Calvin, who argued that the godly life should "bear some resemblance to a fast"[29] and yet that resources should be used not only for necessity but also for "delight and good cheer," austerity is "too severe"; it "degrades" us into "blocks."[30] In praise of frugality, Calvin denounced both excess (extravagance, gluttony, ostentation, greed, prodigality, superfluity, etc.) *and* deficiency. This balance seems essential to the nature of frugality.

This balance, however, does not contradict the choice of voluntary poverty as a calling for some—particularly in identifying with the poor and witnessing against economic injustice. This is a noble vocation.[31] For the rest of us, however, while the calling to frugality does not require sharing in poverty, it still demands sharing in solidarity to eliminate poverty.

Frugality is not a world-denying asceticism that makes some feel competitively righteous—but woefully deprived. It is not the triumph of the spirit over the flesh. It is not the self-mortification and material denigration of some of the saints for the sake of spiritual purification, nor the elitist means of some of the Puritans for proving their divine election. These atomistic and ethereal distortions fail to reflect, as I shall argue, the communitarian and materialistic character of authentic frugality.

Frugality is not a fixed formula for production and consumption. It is not legalism. It does not entail the righteous rigidity and casuistic rules that preoccupy some of the frugal. Instead, frugality is a relative concept, expressing a fittingness to appropriate ends.[32] How much is "enough" depends on what, for whom, and for what purpose. Frugality allows for considerable variation in practices, depending on different needs, tastes, and talents; available resources; social and ecological conditions; and so forth. And surely there are a number of occasions for justifiable indulgence—perhaps a festive frugality.

Still, the relativity of frugality is troublesome. It can be interpreted, as it often was historically, as relative to social rank or status, justifying different standards in proportion to means. Thus, it can be an open invitation to rationalizations of excess—one of the worst examples being the acquisitive Andrew Carnegie interpreting himself as a frugal man and audaciously advocating frugality for the poor![33] Similarly, in Aristotle, nobles with the means for magnificence seem to be exempt from the standards of temperance; they are justified in grand expenditures for great achievements for the common good *and* for themselves.[34] In this context, while we cannot have fixed rules, advocates do need to develop operating principles of frugality—and these principles need to highlight just distribution (socially, intergenerationally, and ecologically) and place a burden of moral justification on luxuries for some in the absence of necessities for all.

Frugality is not a strictly individualistic phenomenon. Many rationales for frugality focus on the spiritual and moral well-being of the individual. The quest of acquisitions, for example, can harm the development of moral character and distract from ultimate obligations to God, even becoming idolatry. This concern is appropriate but insufficient. Indeed, on individualistic assumptions, high levels of commodity consumption can probably be justified, so long as the individual is not physically, morally, or spiritually harmed. But humans do not exist as atomistic entities. We are social and ecological animals, and our moral responsibilities arise from this relational condition. Thus, appropriate levels of production and consumption should be determined not merely individualistically but also—and primarily—relationally. The measure of frugality is solidarity—the moral response to the fact of interdependence, the commitment to the common good, socially and ecologically, nationally and internationally.[35]

Frugality is not the means to prosperity. This is true normatively but, of course, not descriptively in Protestant history. The irony of frugality in Puritanism is that it helped make the pious prosperous! Though the manifest function of frugality was responsible consumption, the latent functions for some were the accumulation of capital and the production of profit—accompanied in some segments of the movement by corrupting dysfunctions, including the perceptions of wealth as the evidence of divine blessing and poverty of divine disfavor, the decline of covenantal solidarity, and the abandonment of frugality itself.[36] The moral purpose of frugality includes capital formation. Though it is not miserliness or hoarding, frugality surely includes saving and investing—reserving and increasing resources for future plans and needs, like educational expenses or retirement. But frugality also has a more comprehensive purpose, which includes just and generous sharing.

Frugality is not a strategy for keeping the poor in their place. This distortion can be found frequently. In mid-seventeenth-century Massachusetts, sumptuary laws, prohibiting displays of luxury by the poor or the parvenu, were designed not to institutionalize frugality but rather to preserve a stable, hierarchical order, to restrict social mobility.[37] In much early Puritan thought, including Calvin,[38] the concept of vocation was linked to social station, the bounds of which were generally divinely decreed and were to be humanly respected. Similarly, in contemporary debates about economic maldistribution, it is often argued that the globalization of North American standards of living would be ecologically disastrous. Yet it is not always noted that it would be ethically discriminatory to tolerate a double track, in which the nations of the South restrict their production and consumption while the nations of the North continue their wanton prosperity.

Frugality, however, as a middle way, is inherently concerned about both overconsumption *and* underconsumption, particularly when the former is a contributing factor to the latter. Its objective is not to keep the poor in their place, but to enable the poor to rise to a new and adequate place. On a planet with scarce resources and a rapidly rising human population, this objective seems to require both floors and ceilings on production and consumption. In general, the substantial reduction of production and consumption in the North may be necessary to enable sufficient production and consumption in the South. The primary moral claim of frugality is directed not at the poor but at the prosperous.

Frugality is not a return to a rustic or pastoral ethos. This way of life can have a moral integrity that warrants commendation—and it usually is frugal. But it is possible and appealing only for a comparatively few.

It cannot be generalized in a nation of 260 million people, let alone on a planet with 6 billion. Nor can frugality be tied to a particular lifestyle; it is a universal norm, applicable to rural, suburban, and urban settings, local and global.

Moreover, this pastoral way of life can generate its own moral corruptions. It is sometimes a fad or affectation of affluent elites. It often expresses an anti-urban bias. It frequently reflects nostalgia, a yearning for the restoration of a passing pattern of life that was considerably less glorious—and for many, far more miserable—than romantically imagined. It often is more a quest for self-reliance in isolation or communes than the self-restraint for the sake of others that is characteristic of frugality. Frugality is the rejection of a new Gilded Age, but it is not thereby the restoration of an old and largely sentimentalized Arcadian Age.

Frugality is not simplicity. In U.S. cultural history, simplicity has been an equivocal concept, often with both moral and aesthetic applications—from plain clothing to unadorned architecture, from good taste to elementary technology, from functional furniture to unpretentious lifestyles, from an appreciation of natural beauties and wonders to a distaste for the ostentatious. For some, like the Quakers, simplicity was an expression of social equality, but for others it was an expression of distinguished gentility. None of these multiple and confused meanings, even though some are admirable, is an appropriate synonym for frugality.

More important, the common identification of simplicity and frugality makes frugality appear irrelevant to modernity. 'Tis not necessarily a gift to be simple; it is a curse when complex problems require complicated solutions. Frugality is not an anti-intellectual phenomenon. For example, both the concept and the practice of frugality are in themselves complex, requiring sophisticated reflection on responsible production and consumption. Nor is frugality an anti-technological phenomenon. It is quite compatible with a recognition of the ambiguities and great variations in the values of technologies. While some forms of technology are socially and ecologically perilous, others, like some medical innovations, can enhance the quality of life and expand human choices. Still others can improve energy and resource efficiency, reduce toxic emissions, and advance ecological knowledge. Frugality offers no wholesale indictment or exoneration of technology. It insists, instead, that ethically acceptable technologies must be "appropriate" to relevant values and social and ecological conditions.[39] "Appropriate," in fact, allows for different scales of technology, from small to large, low to high tech, depending on the situations.

Frugality certainly has been part of a total discipline of piety sometimes called "simplicity" in Christian history.[40] But frugality is not bound to that tradition. Frugality is an economic discipline that can characterize a variety of individuals and even pluralistic societies, apart from particular pieties.

Frugality as the Quest for Abundant Life

The norm of frugality arises in response to several basic questions about the adequacy of material provisions. These questions include: What is a good "quality of life," and what kinds and amounts of goods are necessary or valuable for it? Are various luxuries and conveniences significant benefits or liabilities to personal, communal, and ecological enhancement? How should we distinguish "needs" and "wants"? How much is "enough" in quantity to sustain a reasonable quality of life *and* to ensure that the rest of humanity and other species, present and future, have similar opportunities?

For Christians, the answers to these complex questions depend in part on struggling with the countercultural values encountered in Scripture. They depend, for instance, on dealing with Jesus's radical critique of the idolatries commonly associated with wealth—the idea that one's treasures are indicators of ultimate values (Matthew 6:21). The answers will give heed to the wisdom in Jesus's parables of the rich, hoarding fool (Luke 12:15–21), the rich youth (Matthew 19:16–24; Luke 18:18–25), and the poor, generous widow (Luke 21:1–4), as well as to his disconcerting, hyperbolic teachings about alleviating materialistic anxiety (Luke 12:22–33; Matthew 6:25–33). In this view, there is an insatiable yearning that is morally legitimate in the human creature, but it is not the unquenchable greed for economic goods. Rather, it is the persistent desire for values like love and justice and ultimate meaning that are beyond economic calculations.

Christian answers depend, moreover, on compatibility with the biblical commitments to covenantal justice—which demands provisions for the basic needs and rights of all, including a "preferential option" for the economically vulnerable and powerless, as an expression of loyalty to the Lover of Justice (Pslams 99:4) and as the condition of harmony (shalom) in community (Isaiah 32:17).

Frugality is one strategic response to the above questions that seems consistent with the biblical witness, though it is not the only or most radical Christian response that may be so consistent. But frugality also fits well with a practical and responsible involvement in resolving the socioeconomic and ecological problems of the age.

Frugality is an earth-affirming and enriching norm that delights in the non- and less-consumptive joys of the mind and flesh, especially the enhanced lives for human communities and other creatures that only constrained consumption and production can make possible on a finite planet. It is "sparing" in production and consumption—literally sparing of the scarce resources necessary for human communities and sparing of other species that are both values for themselves and instrumental values for human needs. Frugality minimizes harm to humans and otherkind, enabling thereby a greater thriving of all life. At its best, therefore, frugality can be described paradoxically as hedonistic self-denial, since it is a sensuous concern, or, as Alan Durning notes, "a true materialism that does not just care *about* things, but cares *for* them."[41] From a Christian perspective, frugality is a grateful affirmation of the biophysical in fidelity to the God who lovingly made the biophysical as Creator, who united lovingly with the biophysical as the incarnated Christ, and who dwells lovingly in the biophysical as the sacramentally present Spirit.

This type of frugality seeks the "plenitude" of which Theodore Roszak speaks—the truly abundant life, the "fruitfulness" that the Latin root of *frugality* suggests. It is wanting more, but more of a different kind than economic abundance promises. It is a quest for *being* more rather than *having* more—that is, a qualitative rather than quantitative enrichment.[42] Most subversively in a goutish, overindebted culture, it is consuming *less* than we can afford so that others can have enough. It is not shunning prosperity, but redefining prosperity in less consumptive forms. It is a concern not for the wealth of nations per se, but for the welfare of nations and nature. From the perspective of frugality, wealth has moral significance only insofar as it contributes to just and sustainable welfare—and it seeks to redefine both wealth and welfare in qualitative ways that cannot be simply correlated with aggregate economic indicators like GNP.

Frugality certainly recognizes the great value of economic productivity. That is the foundation of human development. The quality of human life depends on a sufficient quantity of goods and services to satisfy basic biophysical needs and to enable the exercise of our unique creative powers, which, in turn, depends on sufficient kinds and amounts of economic productivity. Frugality is committed to sufficient production, the just distribution of the products, and the reduction of wasteful by-products in order to achieve full human development. It rejects indiscriminate material production, which has made the maximum accumulation of goods an end in itself, and which serves not as a sign of but as a substitute for human well-being. When the economic objective is transformed into an end in itself, rather than the foundational means for human development, human distinctiveness is reduced to being the only mass production mammal and the one with the most voracious appetite.

Thus, the essence of frugality is sacrifice for the sake of higher ends—in fact, for Christians, sacrifice for the sake of Christ's cause of love. This sacrifice is real. Some comforts and pleasures, some lesser values that might be acceptable in an atomistic context but not in a relational one, must be given up. Frugality cannot be made pretty by denying the values lost. Nevertheless, frugality is a form of sacrifice that promises to bring fullness of being in solidarity.

As a norm of moderation, however, frugality does not go as far sacrificially as some might contend the Christian ethic itself demands or as far as some Christians believe that ethic demands vocationally for them.[43] Frugality is clearly not "holy poverty." The temperate character of frugality, however, gives it a distinct advantage over other contenders in the context of moral pluralism. Unlike austerity or other "severe" norms of consumption, which seem to depend on some Christian particularities concerning sacrifice, frugality is on common ground, the "natural law." It is an ethical norm that can be rationally justified and universalized in a morally pluralistic world, apart from appeals to epistemologically privileged revelations or traditions. It is not dependent on a Christian confession, even though that confession is a primary historical source of the norm and provides the fullest warrant for the norm. A coherent case can be made for frugality as a rational and just response to the economics–ecology dilemma. Indeed, that case often is made by contemporary environmental and social analysts—and in ways that can substantially enrich Christian perspectives. Frugality provides a shared standard, perhaps part of a "global ethic," for strategic cooperation among practitioners of various moral traditions.

While the quest for frugality is a strategic response to the initial questions, it does not in itself provide intellectually satisfying answers. A compelling case for frugality depends on comprehensive and coherent answers to such basic questions as the relationship between the quality of life and the quantity of goods. This task seems to be an indispensable part of the mission of ethics in our time.

Frugality for Love, Justice, and Sustainability

Frugality in classical Christian ethical interpretations is an expression of love—seeking the good or well-being of others in response to their needs and to the God who is love. The source of the sacrificial dimension in frugality is love of neighbor, for love always entails giving up at least some of our self-interests and benefits for the sake of the welfare of others in communal relationships.

The connection between frugality and love is common in much-berated Puritanism. It is found in John Winthrop's "A Model of Christian Charity": "In brotherly [affection]," we are called "to abridge ourselves of our superfluities for the supply of others necessities" (though Winthrop is also careful here to remind his compatriots to stay in their stations).[44] It is prominent in John Wesley's sermon "The Use of Money," famous among Methodists for the homiletical maxim: "Gain all you can. Save all you can. Give all you can" (with the emphasis on the last, at least for Wesley).[45] William Penn focuses the connection sharply: "Frugality is good if Liberality be join'd with it. The first is leaving off superfluous Expences; the last bestowing them to the benefit of others that need."[46] For John Calvin, the "rule of moderation" is shaped by the "rule of love," as he spells out in nuanced detail. Frugal self-denial leads to liberal help of neighbors, and both are part of stewardly accountability to God.[47]

These Puritan positions on the frugality–love connection are rooted in a long stewardship tradition of the Catholic church. Indeed, similar comments, sometimes even stronger, can be cited from such Patristic theologians as John Chrysostom, Basil of Caesarea, Gregory of Nyssa, and Ambrose of Milan.[48] For all of these interpreters, frugality ought to be practiced for the sake of just and generous sharing. In the absence of such sharing, frugality is something else: miserliness or hoarding.

In this ethical tradition, distributive justice is one dimension of love; it is love apportioning benefits and burdens, on the basis of relevant similarities and differences, to ensure that all interested parties receive their due or proper share. All human beings, equally and universally in interdependent community, have God-given rights to the material and other essential conditions for expressing their human dignity and for participation in defining and shaping the common good. These rights entail duties of justice by others in proportion to capacities.

This norm contrasts sharply with the present human condition. Radical disparities in economic capacities within and among nations are a fundamental feature of the modern world. These disparities raise important moral questions. If global ecological scarcity is our present or pending condition, is the extensive use of the world's finite resources by the affluent nations a significant factor, actually or potentially, in depriving poor nations of sufficient resources for their essential needs? The linkage between prosperity and poverty is complex and ambiguous. In the context of emerging global interdependence, some factors of prosperity often seem to contribute to poverty,[49] but a causal connection is not always clear or, in some cases, perhaps even present. In this ambiguous context, what are the moral responsibilities of the prosperous to the poor in patterns of consumption?

On the one hand, if a causal relationship exists and if the world's actual and potential material goods are insufficient to tolerate both profligate prosperity for some and economic adequacy for all, then distributive justice requires sufficient reduction of production and consumption of relevant goods in affluent nations in order to make available the material conditions for essential economic and social development in poor nations. In this case, frugality is a *necessary condition* of distributive justice—and, indirectly, of ecological integrity and population stability insofar as these depend on economic sufficiency for all.[50]

On the other hand, the deprivation of necessities for any is an issue of justice and a demand for frugality by the prosperous—whether or not ecological scarcity is a reality, whether or not a causal relationship exists between prosperity and poverty. In all cases, the wealthy must reduce the consumption of their bounty in order to share essentials with and enable sufficiency for the poor. Some might argue, however, that the case for justice here is unconvincing. Instead, the issue is understood as one of benevolent sharing or charitable generosity. If so, that standard is quite demanding in itself, even exceeding the minimal demands of justice! Frugality would then be an expression of love as benevolence.

A similar argument for frugality applies in the case of just responsibilities to future generations or sustainability. Sustainability is living indefinitely within the bounds of the regenerative, assimilative, and carrying capacities of the planet. It is an extension of the covenant of solidarity to the future.

Yet a primary characteristic of present patterns of using the planet as source and sink is *un*sustainability. The moral problem in excessive production and consumption is not only the damage done in the present, but also the harm caused to future generations. A portion of humanity is receiving generous benefits by living beyond planetary means, while future generations will bear most of the risks and costs—from nuclear waste and possible climate change to species' extinctions and soil erosion. In this context, frugality is an essential condition of sustainability. It is the adaptation to biophysical limits for the sake of a just distribution to future generations.

Finally, concern for the just treatment of nonhuman life presents another important argument for frugality—even in the absence of scarcity (though enhanced by it). This concern introduces a major moral limit to economic activity that advocates of unconstrained material growth generally do not recognize or accept. If biophysical scarcity were not a problem for human relations, and if nonhuman lifeforms were nothing but economic instruments—simply "renewable resources" or "capital assets"—to enhance human wealth and welfare, then unconstrained material growth would not be a problem of distributive justice. But if other species are ends for themselves or otherwise moral claimants on moral agents, then obligations are imposed on the human community to restrict production and consumption in order to protect those values. If this second option is true, then economic policies need to pursue the "biocentric optimum" rather than the "anthropocentric optimum" of the first option.[51] On this assumption, profligate production and consumption are anthropocentric abuses of what God has designed for fair and frugal use in a universal covenant of justice.[52] Counseling careful usage and, therefore, minimal harm to otherkind, frugality is the earth-affirming instrument of distributive justice to ensure "enough" scarce goods for all species.

Thus, frugality, as an expression of love, is an indispensable instrument of distributive justice and sustainability. Frugality requires, as the Christian ethical tradition emphasized, *positive duties*—providing goods and services to others from our provisions, such as private gifts or public funds. But it also requires *negative duties*—noninterference through constrained consumption so that enough scarce goods are available for others to make provisions for themselves. Both by positive and negative duties, frugality defines in part the character and conduct of just neighbors.

The Damning Drawback

Frugality is not a major problem for the Sumptuous Society when it is only a personal virtue, even when practiced by several million earnest individuals or thousands of conclaves of the committed. In this case, its social impact is comparatively slight, and even its practical possibilities for individuals are limited by an antifrugal social context.

But if frugality became part of a social ethos, even generating sufficient consent to initiate some structures of coercion, it would likely cause severe economic dysfunctions for the current system. The damning drawback of frugality as a social norm is that it is a formula for market depression in a socioeconomic system that depends on stimulating the economy by encouraging increased consumption. Frugality implies significantly fewer economic stimuli. Thus, it would likely result not only in less production and fewer goods, but also smaller invest-

ments, lower profits, lower wages, reduced revenues, decreased philan-
thropy, and higher unemployment. The social, not to mention the psychological, consequences could be serious, particularly for the poor and
unemployed. Equitable distribution would become an enhanced imperative but probably not an easier task. Internationally, the repercussions
could be no less serious, particularly for poor countries dependent on
affluent countries for export markets and economic assistance. This
superficial analysis of the effects of removing a major component of an
interdependent system indicates the ambiguities of frugality—its long-
term benefits and short-term liabilities. Frugality is an imperative for
sustained human and biotic welfare, but it also could be a source of
human agony under existing conditions. It is an essential condition of
justice but a temporary cause of injustice.

This ironic dilemma does not mean that ethics should forgo frugality as a social norm. The task of ethics is not to adapt reasonable
norms to fit current practices, but rather to challenge and enable societies to adapt their practices to fit those norms. Frugality simply does
not fit the economic assumptions of the Sumptuous Society. Frugality
points to a new economic paradigm that fits this norm, because this
norm fits, as the consumptive model does not, the biophysical limits of
the planet, to which all societies—and norms—must adapt. Frugality
offers the only potentially realistic means of resolving the economics–ecology dilemma.

Yet, for any ethics that claims consequences matter, the advocacy of
frugality as a social norm is irresponsible apart from a concern for its
potentially grievous socioeconomic effects, particularly the dangers of
economic *in*sufficiency and social injustices, and a commitment to envision and advance the policy directions for preventing those effects.
Social frugality need not have the predicted dire effects that opponents
love to recite in generating fear. On the contrary, the simplistic "consume or decline" choice seems to be a "myth."[53] In fact, the opposite
seems to be the case. *Un*sustainable use often means increased *un*employment and reduced production in affected industries, such as is now
occurring in fisheries and forestry in key places. But we need technical
strategies to show the means and ends of economic conversion. These
strategies will need to cover a variety of public policies, including trade
restrictions to protect environments and workers globally and nationally, tax incentives and disincentives to affect supply and demand, public regulations to set limits on products and by-products, and full
employment plans. Indeed, some strategies are already emerging—for
instance, the proposals of the Worldwatch Institute in its annual *State
of the World*, the "steady state" economy of Herman Daly,[54] and the

ecologically "restorative economics" of Paul Hawken.[55] These deserve serious consideration, critical review, and creative responses.

The really perplexing problem, however, is not technical but moral: whether sufficient public consent can be generated to create fair and frugal economies. As a start, this will require a lot of character formation.

Conclusion

This essay points to only the ethical potential and dilemmas of frugality. This norm deserves much greater theoretical development and practical interpretation than it has received in modern Christian ethics. When freed from distortions, frugality seems to be one of the cardinal virtues for our age. As an expression of love, frugality is an indispensable instrument of social and ecological justice and sustainability.

Despite its mild-mannered reputation in ethics, frugality is a radical norm by current social standards—as the deep anxiety it evokes among some of the affluent suggests. It represents a revolt against some basic economic assumptions of the Sumptuous Society. But it is a positive not a negative norm, entailing sacrifices that are personally and relationally enriching. It is an alternative to the American Dream, a competing vision of the future—one that promises fullness of being in solidarity.

The present political prospects for the socialization of frugality are, of course, bleak. But we are not fated. Political possibilities can be transformed, particularly when serious initiatives are undertaken to create a widespread awareness of the values of frugality and the disvalues of alternatives. Contrary to some deterministic distortions of realism, sin is not the only or even the prime feature of the human creature. There is also a strong—though theologically belittled—potential for good, empowered by common grace. Indeed, no social life is possible without a substantial level of human goodness, including a sense of justice and benevolence. An authentic realism must allow for the ambiguous potentialities of both good and evil. No social reform is possible without an acute consciousness of human sin. But equally, no social reform is possible without considerable confidence in human goodness. Hence, the socialization of frugality has limited but still significant potential.

Notes

1. See, for example, the works of the New Road Map Foundation, P.O. Box 15981, Seattle, WA 98115.
2. Two of these rare articles that I find appealing are William E. Gibson, "The Lifestyle of Christian Faithfulness," Dieter T. Hessel, ed., *Beyond*

Survival: Bread and Justice in Christian Perspective (New York: Friendship Press, 1977), pp. 111–40; and Jay B. McDaniel, "Christianity and the Pursuit of Wealth," *Anglican Theological Review* 69, no. 4 (October 1987), pp. 349–61. Both emphasize the importance of interpreting appropriate consumption relationally.

3. Max Weber, *The Protestant Ethic and the Spirit of Capitalism* (New York: Charles Scribner's Sons, 1958), p. 95. Similarly, Ernst Troeltsch called it "ascetic Protestantism" or "secular asceticism" (*The Social Teachings of the Christian Church*, Olive Wyon, trans. [New York: Harper Torchbooks, 1960], p. 814).

4. Wolfgang Schrage, *The Ethics of the New Testament*, David E. Green, trans. (Philadelphia: Fortress Press, 1988), pp. 102–6, 159–60.

5. David E. Shi, *The Simple Life: Plain Living and High Thinking in American Culture* (Oxford: Oxford University Press, 1985), pp. 49, 277.

6. Quoted in Christopher Lasch, *The True and Only Heaven: Progress and Its Critics* (New York: W.W. Norton, 1991), p. 73.

7. *Westminster Dictionary of Christian Ethics,* James F. Childress and John Macquarrie, eds. (Philadelphia: Westminster Press, 1986).

8. *Encyclopedia of the Reformed Faith,* Donald K. McKim, ed. (Louisville, Ky.: Westminster/John Knox Press, 1992).

9. *Economic Justice for All: Pastoral Letter on Catholic Social Teaching and the U.S. Economy* (Washington, D.C.: National Conference of Catholic Bishops, 1986), pars. 75 and 334. The pastoral assumes the necessity of economic growth for social justice (pars. 78–79, 156–58).

10. *Christian Faith and Economic Life: A Study Paper for the United Church of Christ,* Audrey Chapman Smock, ed. (New York: United Church Board for World Ministries, 1987), pars. 223–28, 253–57.

11. Presbyterian Eco-Justice Task Force, *Keeping and Healing the Creation* (Louisville, Ky.: Committee on Social Witness Policy, Presbyterian Church USA, 1989), pp. 72–75.

12. R.H. Tawney, *Religion and the Rise of Capitalism* (Harmondsworth, Great Britain: Pelican Books, 1938), pp. 111–22; David Shi, *The Simple Life*, pp. 8–27.

13. For some refutations of this assumption, see Prentice L. Pemberton and Daniel Finn, *Toward a Christian Economic Ethic: Stewardship and Social Power* (Minneapolis: Winston Press, 1985), pp. 134–36; Robert H. Nelson, *Reaching for Heaven on Earth: The Theological Meaning of Economics* (Lanham, Md.: Rowman and Littlefield, 1991), pp. 5, 7, 282–83; Herman E. Daly and John B. Cobb, Jr., *For the Common Good: Redirecting the Economy Toward Community, the Environment, and a Sustainable Future* (Boston: Beacon Press, 1989), pp. 5–7, 85–96; and M. Douglas Meeks, *God the Economist* (Minneapolis: Fortress Press, 1989), pp. 158–67.

14. See Richard J. Barnet and John Cavanagh, *Global Dreams: Imperial Corporations and the New World Order* (New York: Simon and Schuster, 1994), pp. 163–83.

15. Compare Jorgen Lissner's list of functions outlined in John V. Taylor, *Enough Is Enough* (Minneapolis: Augsburg, 1977), p. 121.

16. Paul L. Wachtel, *The Poverty of Affluence: A Psychological Portrait of the American Way of Life* (New York: Free Press, 1983), pp. 10–16; Juliet B. Schor, *The Overworked American: The Unexpected Decline of Leisure* (New York: Basic Books, 1991), pp. 115–25.

17. Schor, *The Overworked American*, pp. 108, 124–25.

18. Wachtel, *The Poverty of Affluence*, pp. 65–71, 166–64.

19. Schor, *The Overworked American*, p. 134.

20. Daly and Cobb, *For the Common Good*, e.g., pp. 296–97, 374; Herman E. Daly and Kenneth N. Townsend, eds., *Valuing the Earth: Economics, Ecology, Ethics* (Cambridge, Mass.: MIT Press, 1993). See also Cobb's *Sustainability: Economics, Ecology, and Justice* (Maryknoll, N.Y.: Orbis, 1992); William Ophuls with A. Stephen Boyan, Jr., *Ecology and the Politics of Scarcity Revisited: The Unraveling of the American Dream (New York: W.H. Freeman, 1992)*. Christopher Lasch, *The True and Only Heaven: Progress and Its Critics* (New York: W.W. Norton, 1991), esp. pp. 168–69, 530–32. Ian G. Barbour, *Ethics in an Age of Technology* (San Francisco: HarperSanFrancisco, 1993), pp. 251–54, 256–57; Lester W. Milbrath, *Envisioning a Sustainable Society* (Albany: State University of New York, 1989), pp. 218–31; Theodore Roszak, *The Voice of the Earth* (New York: Simon and Schuster, 1992), esp. chap. 9; Erazim Kóhák, *The Embers and the Stars: A Philosophical Inquiry into the Moral Sense of Nature* (Chicago: University of Chicago Press, 1984), esp. p. 108; Schor, *The Overworked American*, pp. 137–38, 164–65; Wachtel, *The Poverty of Affluence*, pp. 48–54. Lester Brown and associates at the Worldwatch Institute. See in general the essays in the annual *State of the World* (New York: W.W. Norton, annual since 1984), esp. Lester R. Brown, Christopher Flavin, and Sandra Postel, "Picturing a Sustainable Society" (1990), pp. 187–80; Brown, "The New World Order" (1991), pp. 18–20; Brown, "Launching the Environmental Revolution" (1992), pp. 174–90; Postel, "Carrying Capacity: Earth's Bottom Line" (1994), pp. 3–21; and Alan Durning's 1991 essay, "Asking How Much Is Enough," pp. 153–69. See also Alan Durning, *How Much Is Enough? The Consumer Society and the Future of the Earth* (New York: W.W. Norton, 1992).

21. Ophuls, *Ecology and the Politics of Scarcity Revisited*, p. 17.

22. One notable exception in which frugality is evident is Douglas John Hall, *Imaging God: Dominion as Stewardship* (Grand Rapids: Eerdmans; New York: Friendship Press, 1986), pp. 1–13, 193–204.

23. Aristotle, *Nicomachean Ethics*, 1.9, 1.2, 7.10; Thomas Aquinas, *Summa theologiae*, I–II, 61.5, 62.4.

24. Aristotle, 3.10–11, 7; Aquinas, I–II, 63.4, 60.5, 61.2–3.

25. Aristotle, 2.6, 7, 9; Aquinas, I–II, 60.1, 4; 65.1.

26. Aquinas, I–II, 60.5; 66.3.

27. On "sustainable sufficiency," see Robert L. Stivers, *Hunger, Technology and Limits to Growth: Christian Responsibility on Three Ethical Issues*

(Minneapolis: Augsburg, 1984), pp. 128–38. See the argument also of William E. Gibson, "The Lifestyle of Christian Faithfulness," in Dieter T. Hessel, ed., *Beyond Survival* pp. 127–36.

28. Theodore Roszak, *The Voice of the Earth*, pp. 254–55.

29. John Calvin, *Institutes of the Christian Religion*, Library of Christian Classics, John T. McNeill, ed.; Ford Lewis Battles, trans. (Philadelphia: Westminster Press, 1960), p. 4.12.18.

30. Calvin, 3.10.2–3

31. See John Cobb, Jr., "Christian Existence in a World of Limits," *Environmental Ethics* I (Summer 1979), pp. 149–58. Also, his *Sustainability*, pp. 13–17, 27–31.

32. Cf. John V. Taylor, *Enough Is Enough* (London: SCM Press, 1976), pp. 45–46.

33. Shi, *The Simple Life*, pp. 35–36, 160–63.

34. Aristotle, 4.2.

35. John Paul II, *On Social Concern [Sollicitudo Rei Socialis]* (1987), pp. 26, 38–39; see also Donal Dorr, "Solidarity and Integral Human Development," in Gregory Baum and Robert Ellsberg, eds., *The Logic of Solidarity* (Maryknoll, N.Y.: Orbis, 1989), pp. 148–53.

36. See especially R. H. Tawney, *Religion and the Rise of Capitalism*.

37. Shi, *The Simple Life*, p. 15.

38. Cf. Calvin, 3.10.6

39. For an appropriately nuanced perspective on technology, see Ian Barbour, *Ethics in an Age of Technology*, pp. 4–25.

40. See Richard J. Foster, *Freedom and Simplicity* (San Francisco: Harper-Collins, 1981). "The Complexity of Simplicity" is the title of his first chapter. Cf. John V. Taylor, *Enough Is Enough*, pp. 80–82.

41. Alan Durning, "Asking How Much Is Enough," p. 169.

42. See John Paul II, *On Social Concern*, p. 28; *On the Anniversary of Rerum Novarum [Centesimus Annus]* (1991), p. 36.

43. See Robert Stivers, *Hunger, Technology and Limits to Growth*, pp. 141–50.

44. Quoted in Robert N. Bellah et al., eds., *Individualism and Commitment in American Life: Readings on the Themes of Habits of the Heart* (New York: Perennial Library, 1987), pp. 22–27.

45. Edward H. Sugden, ed., *John Wesley's Fifty-Three Sermons*, (Nashville: Abingdon, 1983), pp. 632–46.

46. Quotation is from *Fruits of Solitude*, as quoted in J. Philip Wogaman, *Christian Ethics: A Historical Introduction* (Louisville: Westminster/John Knox Press, 1993), p. 142.

47. *Institutes*, 3.10.1–6; 3.7.5; 3.19.7–12.

48. See William J. Walsh and John P. Langan, "Patristic Social Consciousness: The Church and the Poor," in John C. Haughey, ed., *The Faith That Does Justice* (New York: Paulist Press, 1977), pp. 113–51.

49. For example, Northern exploitation of Southern resources, Northern exports of toxic industries and wastes, trade inequities favoring the North,

international debt burdens of Southern nations, the destructive and corrupting impacts of some multinational corporations, and the impoverishing effects in many nations of "structural adjustments" imposed by the World Bank, as well as the heritage of colonialism.

50. These widely recognized connections are a common theme in the World Commission on Environment and Development, *Our Common Future* (New York: Oxford University Press, 1987); also Alan Durning, "Ending Poverty," in Lester R. Brown, ed., *State of the World 1990* (New York: W.W. Norton, 1990), pp. 174–90.

51. See Herman E. Daly, "Elements of Environmental Macroeconomics," in *Sustainable Growth: A Contradiction in Terms?* report of the Visser't Hooft Memorial Consultation (Geneva: Visser't Hooft Endowment Fund for Leadership Development, 1993), p. 47.

52. For a defense, see my *Loving Nature: Ecological Integrity and Christian Responsibility* (Nashville: Abingdon Press, 1991), chap. 6; and my "Biotic Rights and Human Ecological Responsibilities," in Harlan Beckley, ed., *The Annual of the Society of Christian Ethics, 1993* (Boston: SCE, 1993), pp. 137–62.

53. Durning, *How Much Is Enough,* pp. 105–16.

54. Besides *For the Common Good* with John Cobb, see his essays in Daly and Townsend, eds., *Valuing the Earth,* especially his response to criticisms and caricatures of the steady state, pp. 365–81.

55. Paul Hawken, *The Ecology of Commerce: A Declaration of Sustainability* (New York: HarperBusiness, 1993).

Part IV

ETHICS AND PUBLIC POLICY

Overview of Perspectives on Ethics and Policy

Audrey R. Chapman

The impact of current population trends and consumption patterns on sustainability raises a series of significant ethical issues. Sustainability, as defined by James Nash, "is living within the bounds of the regenerative, assimilative, and carrying capacities of the planet indefinitely, as an expression of a covenant of solidarity with future generations."[1] Given the already overstressed state of ecological systems, many scientists view with increasingly alarm the impact of anticipated global population growth and rises in output to meet economic needs and demands during the next half century. They fear that the acceleration of current trends will wreak high, potentially irreversible, damage on the environment. These prospects challenge four fundamental obligations: our responsibility to the Creator for the care and nurture of the creation, recognition of the rights of all people to have the essential material conditions of life, bioresponsibility to other species, and duties to future generations.

At the November 1995 conference, "Consumption, Population and the Environment: Religion and Science Envision Equity for an Altered Creation," there was a consensus that we live at a critical junction in human history. To achieve greater justice and sustainability, the future must be fundamentally different than either the past or the present. Principles of fairness require that all peoples have access to the benefits from and use of the resources of the planet so that they may have the means to life. In an era in which income disparities are growing significantly between the rich and the poor, reducing these inequalities is a major concern. Moreover, chronic poverty affects every dimension of the ecological crisis. Norms of intergenerational equity, as well as religious obligations to be good stewards of God's creation, preclude continuing the depletion of resources and degradation of the environment. Instead we must begin to treat the natural world as a legacy inherited from previous generations and held as a trust for future generations.[2] If we take these norms seriously, the needs of future generations will provide a significant constraint on the consumption of natural resources and on our lifestyles. Similarly, concepts current in environmental ethics, such as the integrity of creation and the intrinsic worth of all species, will mean little if human beings continue to use up planetary resources and impinge on the very ecological systems on which other forms of life are dependent.

On a planet where population growth and economic activity are already causing severe stresses in the natural environment, equity considerations require industrialized countries that use a disproportionate level of resources to reduce their demands on the ecosystem. As various essays in the science section of this volume underscore, population impact is determined not only by total numbers of people, but also and particularly by technologies and living standards. This means that the United States rather than China has the greatest impact of any country on our planet. Because contemporary societies are approaching or have already overshot the optimal scale of human activity, the majority of people who are poor can have greater access to resources, and the patrimony of future generations can be conserved only if the affluent take less.

Thus, a sustainable future means more than global economic growth qualified by environmental sensitivity. It also requires local, regional, and global communities that are "economically viable, socially equitable, and environmentally renewable."[3] One of the most inclusive visions of a sustainable society is contained in a 1974 World Council of Churches statement. Going well beyond the requirements of sustainable development, the World Council of Churches' concept of sustainable society melds social equity, renewable development, and protection of the environment. It has four dimensions:

> *First,* social stability cannot be obtained without an equitable distribution of what is in scarce supply and common opportunity to participate in social decisions. *Second,* a robust global society will not be sustainable unless the need for food is at any time well below the global capacity to supply it, and unless the emissions of pollutants are well below the capacity of the ecosystem to absorb them. *Third,* the new social organization will be sustainable only as long as the rate of use of non-renewable resources does not outrun the increase in resources made available through technological innovation, *Finally,* a sustainable society requires a level of human

activity which is not adversely influenced by the never ending, large and frequent natural variations in global climate.[4]

How can we move from a consumption-oriented society to create a truly sustainable future? A transition to a more sustainable society has many dimensions—personal, social, environmental, economic, moral, and spiritual. It requires both a change in the values, goals, behavior, and lifestyle of people and the establishment of more viable, equitable, and effective economic, social, and political structures. Movement toward these goals is bound up with major changes in lifestyle, economic patterns, and technologies. In turn, this watershed transformation will depend on a revolution in values and goals away from an obsession with material progress and accumulation and toward a greater emphasis on the spiritual and relational dimensions of life. Given the focus of our consumer society, this reorientation may be compared at one level with the experience of overcoming an addiction and at another with the *metanoia,* or turning, associated with religious conversions. Essays in this section explore the ethical and policy dimensions of these issues.

Ethical Norms

The achievement of greater equity and sustainability depends on moral judgments involving questions of desirable ends or living conditions, just distribution, and responsibilities for conserving the creation. It also requires the cultivation of relevant moral virtues and the adoption of an orientation very different from the assumptions at the heart of our consumer society.

At the November 1995 conference, there was also a consensus that consumption patterns need to change and that that will require a revolution in values and expectations to counter the rampant materialism and wastefulness of contemporary affluent societies. David Crocker examines here those issues from the perspective of consumption and well-being. In contrast with the prevalent materialism that identifies well-being with buying, accumulating, and displaying consumer goods, especially those that bring comfort and convenience, he reminds us that commodities fail to give life reliable and ultimate meaning. Steering a middle course between materialism and antimaterialism, he proposes that consumption be evaluated in terms of its ability to promote human well-being. According to Crocker, human well-being requires a balance among four components: physical wellness (being adequately healthy, nourished, clothed, sheltered, and free from physical pain and bodily attack), mental well-being (having the cognitive capacities for perceiving, imagining, reasoning, judging, and deciding), deep personal relationships from which we derive a sense of purpose and self-respect, and opportunities to express our identity.

International Policy Framework

A series of recent United Nations conferences has grappled with population, consumption, and sustainability issues, each developing a final statement or agreement that has contributed to a new international policy framework. The first of these was the 1992 Earth Summit, officially known as the United Nations Conference on Environment and Development, which produced a concluding document known as Agenda 21. Others include the 1994 International Conference on Population and Development, the 1995 Copenhagen Social Summit discussions of equitable development, and the 1996 Fourth World Conference on Women held in Beijing in 1996. Although these conferences usefully highlighted environmental and population concerns and the links between population growth, environmental sustainability, and equitable development, they were less adequate in dealing with the topic of consumption.

Of the conferences, the International Conference on Population and Development in Cairo represents the most significant breakthrough in perspectives about population and development issues. In 1994, the Cairo Conference marked a dramatic shift from current thinking, which conceptualized population goals solely or primarily in a narrow and quantitative focus on numbers of people, demographic goals, and rates of contraception. Instead, the *Programme of Action* of the Cairo Conference emphasizes the empowerment of women, reproductive rights, and improvement in the lives of all people. The final conference statement, a chapter of which is the next piece in this section, recognizes that population growth, development, and environmental factors are interrelated. Perhaps most important, the Cairo Conference gave voice to a new consensus that underscores the need to improve women's condition and to empower women as the most effective way to stabilize population growth and to achieve equitable development. The Cairo consensus puts family planning into the broader context of health, especially women's reproductive health. It includes strong language on women's health rights and the need to improve maternal and child health. There is also strong support for investments in women's and girls' education as a means to reduce fertility and to improve their life opportunities. The document also incorporates text supporting access to safe abortion where it is legal—although it does not endorse abortion as a method of family planning.

While the Cairo consensus represents an important step forward, it is also deficient in several regards. James Martin-Schramm's essay uses the concept of ecojustice to provide a critical moral assessment of the approach to population and consumption issues reflected in the *Programme of Action.* Rooted in the principles of equity and distributive justice, the concept of ecojustice links Christian moral reflection in social ethics and environmental ethics. Martin-Schramm concludes that the Cairo conference represents a major step forward, particularly from the perspective of several ecojustice norms; he raises questions as to whether the means to achieve the stated goals will be sufficient. As he anticipated, there has been a major gap between the rhetoric and the policy and funding commitments to implement them. Another of the weaknesses that he identifies is that the Cairo consensus is far more specific regarding the measures needed to address population growth than it is regarding the measures necessary to deal with unsustainable levels of production and consumption. International protocol requires that the concluding documents from UN conferences be adopted by consensus, and the influence and interests of industrialized countries precluded the adoption of strong sections acknowledging the impact of consumption on environmental sustainability.

Moreover, as might have been expected, there have been noticeable shortfalls in implementing the Cairo global plan. Many countries have yet to summon the political will to carry out their commitments. Five years into the twenty-year plan, there is a massive failure to invest the resources called for in the plan. While spending increased immediately after Cairo, it has since stagnated, in part because of severe economic problems in some countries. By early 1999, the world's richer countries had provided less than one-third of the $5.7 billion they pledged by the year 2000, with the United States likely to be the biggest Cairo deadbeat. In many countries, a lack of health services has made it impossible to offer family planning as part of a broader health care system. This has hampered efforts to reduce maternal and infant deaths. In others, policy changes, such as in the system of distribution of contraceptives, have triggered new problems.[5] Hopefully the global community, motivated by pressures from an increasingly organized and vocal women's network, will do better in the years ahead.

Ethical Dilemmas

As the world population passes the six billion mark, an important question is how we can reduce rates of population growth and do so in a manner consistent with ethical and human rights norms. This concern is the subject of Susan Power Bratton's essay. She points out that far from "population regulation" being only a twentieth-century passion, throughout history human societies have implemented various forms of social sanctions and positive and negative incentives in efforts to suppress or stimulate population growth. Her analysis of biblical narratives related to population growth identifies potential instruction in avoiding coercive and unjust reproductive policies. On the basis of her interpretation of relevant biblical narratives, Bratton determines that positive incentives and reward systems to reduce fertility can be ethically acceptable if the entitlements do not undermine either the religious beliefs or social identity of groups and if the "benefits" are not forced. Negative incentives, by which a person who does not cooperate suffers some harm or loss, tend to be more problematic from an ethical standpoint. Bratton underscores that all coercive methods of encouraging reduced birth rates restrict the rights and choices of individuals. Therefore, she concludes that coercive methods should be employed only where famine is a serious threat or when mortality due to widespread malnutrition is occurring and the crisis is directly related to population levels.

Economic and Policy Issues

Moving toward a sustainable society and economy requires significant national policy changes. Above all else, sustainability requires that U.S. society reduce its overall level of material consumption and waste and choose consumer goods that will have minimal impact on the environment. As Michael Brower and Warren Leon's essay in part 2 states, this norm applies not only to individual Americans but also to business and government and even the U.S. military, which is one of the world's largest sources of waste and pollution. Sustainability also involves ensuring equity in the distribution of environmental impacts and risks among all Americans. According to Brower and Leon, Americans should pay particular attention to three types of consumption—energy, food, and hazardous chemicals—and should reduce waste of all kinds.

How do we get from where we are now to a more sustainable future? As Neva Goodwin describes our situation in her essay, we are in "the transition to a transition." To comprehend the necessary changes, we need to understand our socioeconomy as a coherent, internally consistent whole from which it is hard to alter any one of the interlocking elements without breaking others or distorting the whole pattern. Currently, our socioeconomic system and its values are shaped by the profit motive as a driver of business and the impact of consumerism. According to Goodwin, we have a stark choice—to cling to our current conceptions of growth, without significantly altering our pattern of consumption until environmental or social backlash forces drastic change on us, or to work out a different kind of growth in which the dividend of increased potential output from increasing labor productivity is deployed in ways that strengthen the whole fabric of society. A new, more sustainable paradigm will require a significant reordering of the relationship between work and income, a change in the composition of what is produced, a shift of public and private expenditures and the transfer of jobs out of areas that lack social utility or produce net harm and into areas that combine environmental, social, and personal benefits. A shift in widely held values will be as necessary as a reorientation of economic goals.

What then can provide the motivation to begin the process? Bruce Babbitt, the current U.S. secretary of the interior, offers one potential stimulus to this goal. He notes in his contribution to this volume that one prerequisite for environmentally supportive policy reforms is recognizing that more is at stake than utilitarian values. "It is about a larger issue called humility in front of God's creation."

Notes

1. James Nash, "Population, Environment, and Consumption," in *Interfaith Reflections on Women, Poverty and Population* (Washington, D.C.: Center for Development and Population Activities, 1996), pp. 75–94.
2. Edith Brown Weiss, *In Fairness to Future Generations: International Law, Common Patrimony, and Intergenerational Equity* (Tokyo: United Nations University; Dobbs Ferry, N.Y.: Transnational Publishers, 1989).
3. Larry L. Rasmussen, *Earth Community, Earth Ethics* (Maryknoll, N.Y.: Orbis, 1996), 140–41.
4. This text is quoted in Rasmussen, pp. 138–39.
5. Barbara Vobejda, "Problems Impede Global Plan to Curb Population Growth," *Washington Post,* February 7, 1999, A3.

Report of the International Conference on Population and Development

This is an unedited version of chapter 2, "Principles," of the Programme of Action *of the report of the International Conference on Population and Development held in Cairo, September 5–13, 1994 (New York: United Nations, 1995).*

The implementation of the recommendations contained in the *Programme of Action* is the sovereign right of each country, consistent with national laws and development priorities, with full respect for the various religious and ethical values and cultural backgrounds of its people, and in conformity with universally recognized international human rights.

International cooperation and universal solidarity, guided by the principles of the Charter of the United Nations, and in a spirit of partnership, are crucial in order to improve the quality of life of the peoples of the world.

In addressing the mandate of the International Conference on Population and Development and its overall theme, the interrelationships between population, sustained economic growth and sustainable development, and in their deliberations, the participants were and will continue to be guided by the following set of principles:

Principle 1

All human beings are born free and equal in dignity and rights. Everyone is entitled to all the rights and freedoms set forth in the Universal Declaration of Human Rights, without distinction of any kind, such as race, colour, sex, language, religion, political or other opinion, national or social origin, property, birth or other status. Everyone has the right to life, liberty and security of person.

Principle 2

Human beings are at the centre of concerns for sustainable development. They are entitled to a healthy and productive life in harmony with nature. People are the most important and valuable resource of any nation. Countries should ensure that all individuals are given the opportunity to make the most of their potential. They have the right to an adequate standard of living for themselves and their families, including adequate food, clothing, housing, water and sanitation.

Principle 3

The right to development is a universal and inalienable right and an integral part of fundamental human rights, and the human person is the central subject of development. While development facilitates the enjoyment of all human rights, the lack of development may not be invoked to justify the abridgement of internationally recognized human rights.

The right to development must be fulfilled so as to equitably meet the population, development and environment needs of present and future generations.

Principle 4

Advancing gender equality and equity and the empowerment of women, and the elimination of all kinds of violence against women, and ensuring women's ability to control their own fertility, are cornerstones of population and development-related programmes. The human rights of women and the girl child are an inalienable, integral and indivisible part of universal human rights. The full and equal participation of women in civil, cultural, economic, political and social life, at the national, regional and international levels, and the eradication of all forms of discrimination on grounds of sex, are priority objectives of the international community.

Principle 5

Population-related goals and policies are integral parts of cultural, economic and social development, the principal aim of which is to improve the quality of life of all people.

Principle 6

Sustainable development as a means to ensure human well-being, equitably shared by all people today and in the future, requires that the interrelationships between population, resources, the environment and development should be fully recognized, properly managed and brought into harmonious, dynamic balance. To achieve sustainable development and a higher quality of life for all people, States should reduce and eliminate unsustainable patterns of production and consumption and promote appropriate policies, including population-related policies, in order to meet the needs of current generations without compromising the ability of future generations to meet their own needs.

Principle 7

All States and all people shall cooperate in the essential task of eradicating poverty as an indispensable requirement for sustainable development, in order to decrease the disparities in standards of living and better meet the needs of the majority of the people of the world. The special situation and needs of developing countries, particularly the least developed, shall be given special priority. Countries with economies in transition, as well as all other countries, need to be fully integrated into the world economy.

Principle 8

Everyone has the right to the enjoyment of the highest attainable standard of physical and mental health. States should take all appropriate measures to ensure, on a basis of equality of men and women, universal access to health-care services, including those related to reproductive health care, which includes family planning and sexual health. Reproductive health-care programmes should provide the widest range of services without any form of coercion. All couples and individuals have the basic right to decide freely and responsibly the number and spacing of their children and to have the information, education and means to do so.

Principle 9

The family is the basic unit of society and as such should be strengthened. It is entitled to receive comprehensive protection and support. In different cultural, political and social systems, various forms of the family exist. Marriage must be entered into with the free consent of the intending spouses, and husband and wife should be equal partners.

Principle 10

Everyone has the right to education, which shall be directed to the full development of human resources, and human dignity and potential, with particular attention to women and the girl child. Education should be designed to strengthen respect for human rights and fundamental freedoms, including those relating to population and development. The best interests of the child shall be the guiding principle of those responsible for his or her education and guidance; that responsibility lies in the first place with the parents.

Principle 11

All States and families should give the highest possible priority to children. The child has the right to standards of living adequate for its well-being and the right to the highest attainable standards of health, and the right to education. The child has the right to be cared for, guided and supported by parents, families and society and to be protected by appropriate legislative, administrative, social and educational measures from all forms of physical or mental violence, injury or abuse, neglect or negligent treatment, maltreatment or exploitation, including sale, trafficking, sexual abuse, and trafficking in its organs.

Principle 12

Countries receiving documented migrants should provide proper treatment and adequate social welfare services for them and their families, and should ensure their physical safety and security, bearing in mind the special circumstances and needs of countries, in particular developing countries, attempting to meet these objectives or requirements with regard to undocumented migrants, in conformity with the provisions of relevant conventions and international instruments and documents. Countries should guarantee to all migrants all basic human rights as included in the Universal Declaration of Human Rights.

Principle 13

Everyone has the right to seek and to enjoy in other countries asylum from persecution. States have responsibilities with respect to refugees as set forth in the Geneva Convention on the Status of Refugees and its 1967 Protocol.

Principle 14

In considering the population and development needs of indigenous people, States should recognize and support their identity, culture and interests, and enable them to participate fully in the economic, political and social life of the country, particularly where their health, education and well-being are affected.

Principle 15

Sustained economic growth, in the context of sustainable development, and social progress require that growth be broadly based, offering equal opportunities to all people. All countries should recognize their common but differentiated responsibilities. The developed countries acknowledge the responsibility that they bear in the international pursuit of sustainable development, and should continue to improve their efforts to promote sustained economic growth and to narrow imbalances in a manner that can benefit all countries, particularly the developing countries.

Consumption and Well-Being

David A. Crocker*

Long before many of us began to think philosophically about consumption, we read our children stories that really were parables of consumption, though we did not fully realize this at the time. Some of these stories told of characters who behave foolishly, like the fox hungering for grapes that hang beyond his reach. Others presented characters we might envy for their discernment and good luck. For example, when Goldilocks visits the three bears she has such a sure sense of what is too much, what is not enough, and what is "just right." What is equally fortunate, the right stuff—although it belongs to the three bears and not to her—is readily available. The porridge is at the right temperature; a chair and bed are at the right degree of hardness. And Goldilocks, faced with an array of material goods, unerringly chooses the right thing.

We—consumers in American society—are not usually so lucky. With respect to many consumer goods, we don't know what is too much, what is not enough, and what is just right. Often the right goods are nowhere to be found. When they are available, we frequently fail to choose wisely. Many of us realize that we need a bet-

*David A. Crocker is a senior research scholar at the Institute for Philosophy and Public Policy and the School of Public Affairs at the University of Maryland. This essay was originally printed in *Report from the Institute for Philosophy & Public Policy* 15, no. 4 (Fall 1995): 12–17. An expanded version appears in David A. Crocker and Toby Linden, eds., *Ethics of Consumption: The Good Life, Justice, and Global Stewardship* (Lanham, Md.: Rowman & Littlefield, 1998). The author is indebted to Arthur Evenchik and architect Amanda Martocchio for their comments on an earlier draft.

ter criterion for selection than advertising's image of the good life—if we are to make wise consumption choices, if we are to know what should count as overconsumption, underconsumption, and appropriate consumption.

A given consumption practice may be justifiable or defective in one or more of four ways. First, it may be good or bad for the environment. Some consumer choices deplete scarce resources or damage nature, whereas others contribute to a sustainable way of life. Second, consumption may help or harm other people—our fellow citizens, our descendants, or people in other countries. The benefit or harm I have in mind here is sometimes indirect. For example, in buying moderately priced and fashionable sportswear, we may be a factor in the existence of sweatshops in our own country and abroad; in devoting much or all of our income to personal comforts, we may neglect to assist others less fortunate than ourselves. Third, our consumption practices may affirm or undermine values and institutions deemed essential to our community. Widespread choice of private schooling, for example, may weaken public education and social equality. Deciding to buy a house in the inner city rather than in a suburban neighborhood may strengthen urban institutions aspiring to cultural and class diversity. Finally, a consumption choice or pattern may be beneficial or detrimental to a person's own well-being—apart from its effect on institutions, other people, and the natural world.

My main purpose here is to investigate this fourth way of assessing consumption. What role should goods and services play if our lives are to go well? What kinds of consumption are good for us? Bad for us? What evaluative criteria should we employ to assess the impact on our lives of our present consumption and to evaluate alternative consumption patterns and ways of living?

The consumption norm I offer here will supply general rather than detailed guidance. After all, as Stanley Lebergott has remarked, no principle can tell us that five compact discs, say, is the right number to buy each year, while six is too many and four is too few.[1] Any norm that presumed to set such limits would be arbitrary, even dictatorial. What we require are general principles that will allow each of us to make choices appropriate to the distinctive character of our individual circumstances.

Materialism and Antimaterialism

One way to arrive at a reasonable consumption norm is to assess widely held normative outlooks about the acquisition and possession of commodities and, more generally, about human well-being. Materialism and antimaterialism are two such perspectives. Getting clear on where and how these rival norms go wrong will help us arrive at a better conception of well-being and a more adequate consumption norm.

Materialism assures us that having more is being more; it identifies well-being with buying, accumulating, and displaying consumer goods, especially those that bring comfort and convenience. In our own consumer society, materialism is often perceived as a national characteristic. Though we poke fun at our materialist obsessions—"I SHOP, THEREFORE I AM!" "NOTHING SUCCEEDS LIKE EXCESS!" "SO MANY MALLS, AND SO LITTLE TRUNK SPACE!"—we do not renounce them. According to Juliet Schor, "Americans spend three to four times as many hours a year shopping as their counterparts in Western European countries. Once a purely utilitarian chore, shopping has been elevated to the status of a national passion."[2] In his poem "The Return," Frederick Turner captures the consumerist nostalgia of American soldiers in Vietnam:

> What we miss
> are the bourgeois trivia of capitalism:
> the smell of a new house, fresh drywall, resin
> adhesive, vinyl, new hammered studs; ground coffee
> in a friend's apartment in San Francisco, the
> first day of the trip; the crisp upholstery
> of a new car
>
> It is the things
> money *can* buy we remember, the innocence
> of our unfallen materialism.[3]

The poem looks gently upon the "bourgeois trivia of capitalism" and the wistful soldiers who recall them. But in other contexts, it is harder to see American materialism as innocent or "unfallen." Consider the consumerist manifesto that retailing analyst Victor Lebow issued in 1955 for an American economy enjoying a postwar boom:

> Our enormously productive economy . . . demands that we make consumption our way of life, that we convert the buying and use of goods into rituals, that we seek our spiritual satisfaction, our ego satisfaction, in consumption. . . . The very meaning and significance of our lives is today expressed in terms of consumption. We need things consumed, burned up, worn out, replaced, and discarded at an ever increasing rate.[4]

Although the "buying and use" of commodities may be essential to one kind of economic growth, most of us believe that commodities *by themselves* fail to give life reliable and ultimate meaning. As political economist Robert E. Lane suggests, it is not what we buy or own that brings us happiness but rather our work, our relations with our spouses and colleagues, and the well-being of our children.[5] Indeed, the world of consumer goods, and a life devoted to their pursuit, may insulate us from deeper challenges and human connections.

Antimaterialism, whether religious or nonreligious, feeds on the very real weaknesses of consumerist materialism. Antimaterialists conceive the good life precisely so as to protect themselves from disappointments in the changeable, frustrating world of bodily appetites and worldly possessions. They strive to free themselves of all attachments to material goods, or at least to reduce significantly their level of material consumption. In its most extreme forms, antimaterialism forsakes the world in order to lay up "treasures in heaven," where "neither moth nor rust doth corrupt, and where thieves do not break through nor steal" (Matthew 6:20, KJV). More moderate forms affirm the ultimate importance of this-worldly, but still nonmaterial, realities. Some prize inner rationality, self-possession, and self-sufficiency (as in Stoicism). Others emphasize the fulfillment that comes through personal relationships.

Yet these various ideals of independence from the material world may be just as misguided as the materialist effort to elevate the acquisition and enjoyment of worldly goods to life's supreme aim. One obvious concern is that there is a physical aspect of human well-being, a requirement of certain goods and services that meet basic needs—adequate food, clothing, shelter, health care, and so on. The antimaterialist might concede this point and allow for the modest satisfaction of those needs. Still, the resulting ethic might be so austere that it condemns much that makes life worth living.

It is true that we must avoid being obsessed with or possessed by commodities. Yet we must also honestly recognize the positive role some goods and services play in our lives; otherwise, we risk adopting a critique of consumerism that is blindly at odds with our own choices and experiences. Nutritious and tastefully prepared food consumed with others can be good for both body and soul. Aesthetically attractive dwellings and clothes enable us to shape and express who we are. Marriage rites include the exchange of rings. Although we do well to avoid using presents to manipulate people, we sometimes express parental love and nurture friendships by carefully selected material gifts. We often exercise our civic responsibilities through phone, fax, and email. Air travel—for example, to consumption conferences—brings new ideas and new friends into our lives.

Rather than seek the good life by withdrawing to a self-sufficient inner or transcendent world, many of us, when we are honest with ourselves, believe that we may realize our well-being when we satisfy *certain* worldly desires and utilize *certain* material means. As we meet human needs, realize our best potentials, press against limits, and cope with bad fortune in humanly excellent ways, commodities can play an important instrumental role.

Well-Being

Although they contain some truth, both materialism and antimaterialism are guilty of exaggeration. Those who endorse one of these views are typically engaged in an overreaction against the other. Sometimes entire cultures vacillate between the two, like a car that uncontrollably fishtails from side to side. If we are to have a reasonable consumption norm, the pair must be rejected together and replaced by a balanced and stable conception of the sources and meaning of well-being.

The conception I will present here derives largely from Aristotle's ethic of human flourishing,[6] and from the work of two contemporary philosophers, Amartya Sen and Martha Nussbaum,[7] who have acknowledged their own debt to Aristotle in formulating what is known as the *capabilities approach*. It also owes much to Partha Dasgupta, Thomas Scanlon, and James Griffin, whose orientations are closer to the Kantian or utilitarian traditions.[8]

According to this conception, well-being refers not to one component of life, such as pleasure or the satisfaction of basic needs, but to a heterogeneous list of human conditions, activities, inner capacities, and external opportunities. To have well-being, to be and to do well, is *to function* and *to be capable of functioning* in certain humanly good ways.

The bodily components of well-being include being adequately healthy, nourished, clothed, sheltered, and mobile, as well as being free from physical pain and bodily attack. (The criterion of "adequacy," as applied to clothing and shelter, includes being able to appear in public and not be shamed by one's physical neediness.) In some circumstances, it is true, individuals willingly sacrifice some aspect of bodily well-being, as when a hunger striker allows herself to become malnourished. Such a person forgoes a component of healthy functioning. Notice, however, that she is not thereby *incapable* of being well nourished. She is better off than someone who cannot acquire food or who, on account of illness, is unable to derive nourishment from the food he consumes. Thus, we can say that physical well-being includes certain capacities and opportunities as well as functionings.

Although physical wellness is necessary for well-being, it is not sufficient. We must also include mental capabilities and functions in our conception of well-being. Among these are the cognitive capacities for and activities of perceiving, imagining, reasoning, judging, and deciding. The latter embraces our being able to choose a conception of the good life. Mental well-being also includes opportunities and capacities for enjoying or finding pleasure—whether in other aspects of well-being, such as physical health, or in such things as art and nature. Happiness, although contributory to well-being, is not sufficient, for it may occur with and even camouflage significant deprivation. The severely destitute may expect little of the world and be overjoyed by a "small mercy."[9] Drug-induced euphoria is compatible with morbidity and malnutrition.

Human well-being has a social as well as physical and mental dimensions. We believe we are less than fully human if we lack the deep personal relations of family and friendship as well as participation in wider social and ecological communities. It is largely from these relations that we derive a sense of purpose and self-respect.

A fourth aspect of well-being is what Nussbaum calls strong "separateness" but might be better termed "singularity": "Being able to live one's own life in one's own surroundings and context."[10] In addition to social relations, well-being depends on our being distinct from others, expressing our singular identity, and having that which is uniquely our own.

What is the proper relation between these aspects of well-being? I would argue that the good life requires achieving a certain *balance* among them. Although a particular consumption choice may contribute more to one aspect of well-being than to the other, the wise consumer strives for an overall consumption pattern that promotes balance and harmony among the components. Too much or too little of a good thing in one dimension may decrease our overall well-being in one or more of the others. The person obsessed with physical fitness will have little opportunity or capability for intellectual and social activity. The intellectual's books and the hacker's computer may stunt their owners' physical and social development and prevent full flourishing. My private possessions may distract me from civic participation. Within and among each dimension, the wise consumer avoids the extremes of excess and deficit and seeks moderation.

The ideal of balance has an additional application in our understanding of well-being. We do well to balance the times of our lives. This means, on the one hand, not unduly sacrificing present well-being for our future good, and, on the other, not choosing to obtain certain aspects of well-being now when the likely long-term cost is significant loss of important capabilities.

It might be objected that some "one-sided" lives can still go well or be well chosen. What of the connoisseur whose single-minded devotion to acquisition results in poor health or failed friendships, but who assembles an art collection of unquestioned importance or beauty? What of the musical prodigy who willingly relinquishes some of the pleasures of childhood in order to cultivate her gifts?

There are, I think, several answers to this challenge. One may concede that well-being is not everything. A person's life may be well chosen and worthwhile precisely because he has sacrificed his well-being, or indeed his life, for a noble cause—clan, country, or culture. But even as we honor those who embrace a higher good at the cost of their well-being, we usually regret that such a choice was required. It is also worth saying that absolute choices between different aspects of well-being are less often called for than we might think. Sometimes, for instance, we may be able to *emphasize* one good over another without *eliminating* the subordinated good from our lives. Similarly, as I suggested above, we may emphasize different aspects of well-being at different times. Wynton Marsalis explains that after many years of playing exemplary jazz together, his ensemble disbanded so that its members—now married with families—could spend more time at home.

Unfortunately, when consumption choices do disrupt the balance of our lives and lead to a sacrifice of well-being, the sacrifice often arises not for the sake of exalted achievement or another demanding good, but instead from our own shortsightedness or self-deception. In such cases, we may reasonably conclude that the value of a life of sustained and balanced well-being exceeds the benefits of a consumption choice narrowly devoted to present well-being.

Employing the Norm

The conception of well-being I have described provides the basis for a general consumption norm: One consumption pattern or choice is better than another if it does better in protecting and promoting a person's well-being. A consumption pattern or choice can be criticized to the extent that it weakens or destroys those capabilities and functionings that are the components of well-being.

Such a consumption norm has sufficient content to rule out the one-sidedness of materialism and antimaterialism, and sufficient generality to permit quite diverse "balancing acts," depending on a person's specific abilities, opportunities, and choices. What promotes, maintains, balances, weakens, or destroys the same aspects of well-being can and often does vary from person to person as well as from society to society. Some people can possess more commodities than others before such possession undermines their bodily health, practical rationality, or citizenship by fostering the vices of imprudence, greed, and political indifference. To live well in an opulent, technologically advanced community requires different goods from those required to live well in a poor and traditional one. As Amartya Sen explains:

> To lead a life without shame, to be able to visit and entertain one's friends, to keep track of what is going on and what others are talking about, and so on, requires a more expensive bundle of goods and services in a society that is generally richer, and in which most people have, say, means of transport, affluent clothing, radios or television sets, etc. Thus, some of the same capabilities (relevant for a "minimum" level of living) require more real income and opulence in the form of commodity possession in a richer society than in poorer ones. The same absolute levels of capabilities may thus have a greater relative need for incomes (and commodities).[11]

Wise consumption requires knowledge of ourselves and our society as well as choice in the light of that knowledge.

Given its person- and context-sensitivity, what sort of guidance can a norm derived from the capabilities approach yield with respect to our consumption choices? Certainly we do not need a consumption norm to know that some consumption choices are generally bad for us (high-fat diets, for instance) and others (certain lifesaving medicines, for example) typically good for us. To illustrate the salutary prescriptive force of the capabilities norm, let us suggest its application to the question of housing. Without presuming to give utterly novel housing guidelines, the capabilities norm of well-being will be successful if it clarifies and endorses one strand of our often contradictory everyday judgments about humanely good housing.

Although frequently neglected in the consumption debate, our choice of a dwelling is important, for housing requires a far greater proportion of consumer dollars (31 percent) than any other major category of consumer expenditure. Many people, of course, have no choice but to settle for housing that is clearly at odds with even a modicum of well-being. Others could afford decent housing, but only if they changed their conception of the good life and their overall consumption pattern. Let's suppose, however, that an individual or family has or receives the resources to acquire a dwelling that protects and promotes the four aspects of well-being. What would be the *general* characteristics—compatible with individual and societal variation—of such housing?

First, the capabilities approach requires that housing options should be assessed with respect to their occupants' physical well-being. The neighborhood should be reasonably free of crime as well as of the hazards posed by polluted air and water. Safe, accessible parks and playgrounds should offer opportunities for recreation. The dwelling itself should enable its occupants to be secure from the elements. Physical health requires good ventilation, sanitation, and sunlight as well as adequate space for sleeping, meal preparation, and personal hygiene.

"Livable" housing also protects and promotes the mental component of well-being. In its design and furnishings, good housing occasions aesthetic enjoyment. Maintaining and improving one's housing afford opportunities for the exercise of practical rationality.

Further, good housing safeguards and nurtures various forms of sociability and mutuality. Decent housing provides its inhabitants and their guests with opportunities for conviviality. To permit and encourage wider social participation, good housing is reasonably close to neighbors, work, schools, and cultural opportunities.

Finally, good housing expresses—through such things as workbenches or studies, gardens or basketball hoops—the inhabitants' specific and perhaps distinctive ideas of the good life. Moreover, a good dwelling is sufficiently commodious to provide each occupant with the personal space (and time) that is needed to be able to live one's distinctive life in one's own ambiance. Such "separateness" may be best expressed by each occupant's having his or her own room or part of a room.

Sometimes two or more components of well-being call for the same housing site or structure. A room with good sunlight, for example, can be both healthy and aesthetically pleasing. A shared bedroom can promote both mutuality and singularity. It will often prove difficult, however, to find housing that satisfies (equally) each component of this complex conception of well-being, for the elements of well-being can conflict with as well as support one another. There may be no neighborhood, for instance, that is both close to one's work and reasonably safe or near good schools. Moreover, someone might find housing that would directly secure her well-being but would cost so much that she could not afford necessary food and health care. We must therefore employ practical rationality in order to address the advantages and disadvantages of each specific option, deliberate, and finally judge which abode (and larger consumer pattern) is, on balance, best for us.

In order to focus on the effects of consumption on our own well-being, I began this essay by setting aside other important questions about consumption choices: those involving our moral obligations toward the environment, societal institutions, and other people. In the course of the argument, I have suggested that the pursuit of higher goods does not necessarily require the sacrifice of well-being. Nonetheless, it is true that consumption choices that are ostensibly good for us often harm nature, society, or others and that the norm of our own well-being is not sufficient for assessing these choices. Some of the materials for such an assessment may be found in other essays in this volume.[12]

Notes

1. Stanley Lebergott, "Pursuing Consumption Limits" (paper presented to the conference on "Consumption, Global Stewardship, and the Good Life," Institute for Philosophy and Public Policy, University of Maryland, College Park, September 29–October 2, 1994): 2. See also Stanley Lebergott, "What Principle?" *Reports from the Institute for Philosophy and Public Policy* 15 (Fall 1995): 22.
2. Juliet B. Schor, *The Overworked American: The Unexpected Decline of Leisure* (Basic Books, 1992), 107.
3. Frederick Turner, *The Return*, with a preface by George Steiner (Woodstock, Vt.: Countryman Press, 1981), 5–6, quoted in Chandra Mukerji, *From Graven Images: Patterns of Modern Materialism* (New York: Columbia University Press, 1983), xv.
4. Victor Lebow, "Price Competition in 1955," *Journal of Retailing* 31 (Spring 1955): 7. Quoted in Alan Thein Durning, *How Much Is Enough: The Consumer Society and the Future of the Earth* (New York: W.W. Norton, 1992), 21–22.
5. Robert E. Lane, "Does Money Buy Happiness?" *Public Interest* 111 (Fall 1993): 59, 61.
6. Aristotle, *Nicomachean Ethics,* trans. Terence Irwin (Indianapolis, Ind.: Hackett, 1985).
7. See, for example, Martha C. Nussbaum and Amartya Sen, eds., *The Quality of Life* (Oxford: Clarendon Press, 1993); Martha Nussbaum, "The Good Discipline, the Good as Freedom," in David A. Crocker and Toby Linden, eds., *Ethics of Consumption: The Good Life, Justice, and Global Stewardship* (Lanham, Md.: Rowman & Littlefield, 1998), 312–41; and Amartya Sen, "The Living Standard," in Crocker and Linden, eds., *Ethics of Consumption,* 287–311.
8. T. M. Scanlon, "Value, Desire, and Quality of Life," in Nussbaum and Sen,

eds., *The Quality of Life*, 185–200; T. M. Scanlon, "The Moral Basis of Interpersonal Comparisons," in Jon Elster and John Roemer, eds., *Interpersonal Comparisons of Well-Being* (Cambridge, England: Cambridge University Press, 1991), 17–44; and James Griffen, *Well-Being: Its Meaning, Measurement, and Moral Importance* (Oxford: Clarendon Press, 1986).

9. Amartya Sen, *Resources, Values, and Development* (Cambridge, Mass.: Harvard University Press, 1984), 512.

10. Nussbaum, "The Good as Discipline," 320.

11. Sen, "The Living Standard," 298.

12. See also Crocker and Linden, eds., *Ethics of Consumption.*

Population, Consumption, and Eco-Justice: A Moral Assessment of the 1994 United Nations International Conference on Population and Development

James B. Martin-Schramm*

This essay uses the nascent concept of eco-justice to provide a critical moral assessment of the approach to population and consumption issues reflected in the *Programme of Action* adopted at the 1994 United Nations International Conference on Population and Development (ICPD) in Cairo, Egypt.[1]

The term *eco-justice* emerged in the early 1970s and was coined by the Board of National Ministries of the American Baptist Churches.[2] The concept emerged out of a desire to link Christian moral reflection in social ethics and environmental ethics. Rooted in the principles of equity and distributive justice, eco-justice is increasingly described in relationship to four specific moral norms. These norms are sustainability, sufficiency, participation, and solidarity. The World Council of Churches brought renewed attention to these norms, which are rooted in the Bible and have been developed in Christian traditions, through the themes of its last three world assemblies and related conferences.[3] Recently, the Presbyterian Church (USA) and the Evangelical Lutheran

*James B. Martin-Schramm is an assistant professor of religion at Luther College in Decorah, Iowa. This essay is an edited version of the final chapter of his doctoral dissertation, entitled *Population, Consumption, and Eco-Justice: Challenges for Christian Conceptions of Environmental Ethics*; it was presented at the annual meeting of the Society of Christian Ethics, Albuquerque, New Mexico, January 5, 1996.

Church in America incorporated these norms into policy documents and social statements of their respective churches.[4] Both churches highlight the relation of these norms to the principles of equity and distributive justice, and it is fair to say that these four norms provide the most substantial foundation to date for an ethic of eco-justice.[5]

Accepted by all 179 nations and the six observers that attended the ICPD, the *Programme of Action* represents a significant change from previous attempts to address population and development issues in the international community.[6] As a foundation for guiding population policies for the next twenty years, this new approach represents a shift away from a narrow and *quantitative* focus on numbers of people, demographic goals, and rates of contraceptive use toward a broad and *qualitative* emphasis on the empowerment of women, reproductive rights, and improvement in the lives of all people.

The 1994 *Programme of Action* comes closer to reflecting an eco-justice perspective than have prior national and international policies on population and development. For example, it reflects some awareness of the reciprocal relationship between ecological integrity and social justice. One way in which this is expressed is through its emphasis on the ecological impact of production and consumption levels in the north and population growth in the south. Another way in which this eco-justice perspective emerges is through the document's emphasis on the need to eradicate poverty and improve the quality of life for all people as a vital step on the path to sustainable development. Finally, a third expression of an eco-justice perspective is found in the document's confident affirmation of the empowerment of women and in the various sections that call for the dismantling of patriarchal barriers that prevent their full participation in society.

Despite the important improvements that the Cairo consensus represents, however, there are some significant ways in which the document falls short, from the perspective of an eco-justice critique. For example, the *Programme of Action* gives insufficient attention to challenges posed by overconsumption; its funding targets do not reflect many of its goals; and there is reason to question the depth of political resolve of the signatories of this nonbinding agreement. This essay evaluates these weaknesses along with the strengths of the new approach to population and consumption issues that were agreed upon in Cairo.

For the sake of clarity and brevity, however, it is necessary first to identify those things that this essay does not set out to accomplish. First, it does not attempt to provide a substantial history of the important developments that took place before the conference during the preparatory process, nor does it offer a detailed summary of the daily debates that took place in Cairo. Among other things, such an endeavor would emphasize the important role that women's groups and non-governmental organizations played in shaping the draft *Programme of Action,* the return of the United States to a position of international leadership in the population field, and the response of the Vatican and conservative Islamic countries to various issues.[7]

Second, because this essay focuses on what is unique about the 1994 *Programme,* it does not offer a substantial assessment of less controversial sections dealing with the structure and distribution of populations, technology and research, partnership with the nongovernmental sector, the need for international cooperation, and follow-up to the conference.

Finally, while this essay identifies several moral issues that were discussed at the ICPD, it does not attempt to resolve them in a definitive way. A separate essay would be required to address adequately various controversies such as those revolving around definitions of the family; sexuality within and outside the bonds of marriage; the role of abortion in family planning and reproductive health care; the relationship of adolescent and parental rights; the clash between the concept of reproductive rights and the conservative religious traditions and cultures; and the debate regarding family reunification for migrants and refugees.

What this essay does focus on are the unique features and key objectives of the *Programme of Action.* It begins with a brief historical summary of international reflection on the relationship of population and development. From there it identifies the key objectives of the *Programme of Action* in relationship to the plan's relevant chapters. With this foundation and introduction to the document, I will then offer my moral appraisal of this approach from an eco-justice perspective.

Historical Context

The United Nations Fund for Family Planning Activities (UNFPA) was established by the UN in 1969 to assist all nations that wanted help in studying the size of their populations, evaluating the consequences of population trends, and developing population policies and programs to control fertility.[8] Apart from the assistance UNFPA has offered in these areas, perhaps its most important contribution has been its creation of an international forum for debate concerning population policies and their place in the process of development.

The first UNFPA World Population Conference was held in Bucharest, Romania, in August 1974. The conference served as an important milestone in the history of efforts to "control" global population growth. At the conference, representatives of less developed nations repudiated vigorous family planning programs proposed by developed nations like the United States as simple-minded, racist, and imperialistic attempts to avoid the underlying cause of population growth, which they regarded as poverty and an unjust distribution of resources. Led by China and India, the Southern countries summarized their position with the slogan "Development is the best contraceptive."[9]

By the time the second UN decennial conference took place in Mexico City in 1984, the views of nations in the south and in the north had gone through some important changes. After a decade of rapid population growth, many developing nations recognized the need to include family planning within a comprehensive approach to social and economic development. With the election of the Reagan administration, however, the United States had switched its position. Arguing that demographic concerns were a "neutral factor" in development efforts, the U.S. delegation embraced the conclusion of the 1974 conference that economic development should serve as the principal means to lower the rate of population growth. The leader of the U.S. delegation also announced that U.S. support for family planning programs would be restricted in three ways:

> First, where U.S. funds are attributed to nations which support abortion with other funds, the U.S. contribution will be placed into segregated accounts which cannot be used for abortion; second, the U.S. will no longer contribute to separate non-governmental organizations which perform or actively promote abortion as a method of family planning in other nations; and third, before the U.S. will contribute funds to the UNFPA, it will first require concrete assurances that the UNFPA is not engaged in, and does not provide funding for, abortion or coercive family planning programs.[10]

Conference delegates, who over the previous ten years had seen the results of effective family planning and had grown in their appreciation for its implementation, were stunned. After the conference, the International Planned Parenthood Federation refused to follow the U.S. policy and as a result lost all U.S. funding in late 1984. In 1986, citing the coercive nature of China's family planning program, the Reagan administration removed all funding from UNFPA for its alleged participation in the program.[11]

Given this history, it is all the more remarkable that the third UN decennial conference on population and development in 1994 adopted the approach and achieved the level of consensus that it did. Three important factors contributed to the success of the ICPD. The first factor involved the perspectives of nations in the south. The 1980s were another disappointing decade for most of these nations, as they saw significant efforts at social and economic development substantially mitigated by debt burdens, reduced levels of foreign aid, structural adjustment programs, wars, and continued rapid population growth. Recognizing the relationship between population growth, poverty, and environmental degradation, nations in the south were nevertheless concerned that the world's ecological woes would not be placed on the backs of the poor. Throughout the eighteen-month preparatory process that led up to the ICPD, these nations emphasized that the vast majority of global environmental degradation is the consequence of production and consumption patterns in developed nations.

At the same time, leadership had changed in the United States, and the Clinton administration almost immediately reversed the policies of its predecessors. As the preparations for the 1994 ICPD got under way in 1992, U.S. participation under the leadership of Under-Secretary of State Timothy Wirth played an important role in the development of the draft *Programme of Action*. Together with many other countries, the United States invited the participation and welcomed the input of many nongovernmental organizations in drafting the *Programme of Action* during the preparatory process. Since population and development funds have increasingly been channeled toward these organizations, they had particular expertise to contribute. Approximately half of the United States' official delegation to the ICPD was composed of delegates from nongovernmental organizations.

Among these organizations, none was more important than various women's groups from around the world that had mobilized and effectively strategized for this occasion. Organized in networks that united women from developing and developed nations, these groups emphasized the need to place the welfare of women at the centerpiece of any approach to population and development issues. Throughout the preparatory process and in Cairo, women's groups ably advocated and defended those provisions of the *Programme of Action* that now distinguish it from prior attempts to address those issues.[12]

Summary of Key Features of the *Programme of Action*

The *Programme of Action* approved in Cairo at the conclusion of the ICPD in September 1994 represents a watershed in the history of international reflection on population issues for two main reasons. First, it insists that efforts to address global population growth must be part of the broader goal of achieving sustainable development and an improved quality of life for all. The document emphasizes that people must become the subjects and not the objects of population and development policies. It also stresses that economic growth must take place within the context of sustainable development. Second, the *Programme of Action* emphasizes that the best path to population stabilization is by empowering women and improving their status in society. While the plan reaffirms the goals of predecessor documents regarding universal access to family planning, increases in primary and secondary education, and reductions in infant and maternal mortality, it places these tasks within the broader goals of gender equity, reproductive rights, and the elimination of all discrimination against women and girl children. To gain a better understanding of these new directions, it is necessary to summarize the contents of several key chapters in the *Programme of Action.*

Chapter 3 focuses on the relationship of economic growth to sustainable development. Arguing that development is a "universal and inalienable right and an integral part of fundamental human rights," the chapter emphasizes the need to integrate population and development strategies toward the goal of meeting the needs and improving the quality of life of present and future generations.[13] The chapter gives particular attention to the need for poverty eradication, economic growth within the context of sustainable development, and a concern for women as the poorest of the poor. It calls for increased investment in education, sanitation, housing, job creation, food security, and other services that will promote the development of human resources. It also calls for increases in reproductive health services and the elimination of various barriers to women's full participation in society. Addressing environmental concerns, the chapter emphasizes the need to reduce unsustainable consumption and production patterns as well as negative impacts of demographic factors on the environment.

While the emphasis on gender equity and reproductive rights is found throughout the document, chapters 4 and 7 give specific attention to these key features of the *Programme.* Entitled "Gender Equality, Equity and Empowerment of Women," chapter 4 begins by emphasizing that the empowerment of women is not only a "highly important end in itself, it is also essential for the achievement of sustainable development."[14] Stressing the crucial role that education plays in the empowerment of women, the chapter draws particular attention to studies indicating that of the 960 million people who are illiterate in the world, two-thirds are women. Also, of the 130 million children not enrolled in primary school, 70 percent are girls.[15] In response, the chapter calls for equality and equity between men and women, the full participation of women in all facets of life, and recognition of the need to ensure that all women, as well as men, receive the education they require to meet their needs and exercise their human rights. Toward this end, the chapter calls for the establishment of various mechanisms to facilitate women's equal participation in society, increased efforts to eradicate poverty and illiteracy, and the elimination of all forms of discrimination and violence against women and girl children.

Chapter 7 provides the definitions of reproductive health and reproductive rights and elaborates on these concepts that are so significant in the *Programme of Action*. The chapter defines reproductive health as "a state of complete physical, mental and social well-being and not merely the absence of disease or infirmity, in all matters relating to the reproductive system and to its functions and processes."[16] Reproductive rights are defined in the following way:

> Bearing in mind the above definition, reproductive rights embrace certain human rights that are already recognized in national laws, international human rights documents and other consensus documents. These rights rest on the recognition of the basic right of all couples and individuals to decide freely and responsibly the number, spacing and timing of their children and to have the information and means to do so, and the right to attain the highest standard of sexual and reproductive health. It also includes their right to make decisions concerning reproduction free of discrimination, coercion and violence, as expressed in human rights documents. In the exercise of this right, they should take into account the needs of their living and future children and their responsibilities towards the community.[17]

Affirming and enabling these rights, the *Programme of Action* calls upon all nations to make a full range of comprehensive and voluntary reproductive health-care services accessible through the primary health care system to "all individuals of appropriate ages as soon as possible and no later than the year 2015."[18]

Emphasizing that all people have the right to decide freely and responsibly the number and spacing of their children, and that this requires access to the information and means to do so, the chapter gives particular attention to the unmet demand for family planning services around the world. While 55 percent of all couples in the developing world now use some method of family planning, surveys indicate that at least 350 million couples around the world still do not have access to a full range of family planning methods. Other studies suggest that "approximately 120 million *additional* women would be currently using a modern family planning method if more accurate information and affordable services were easily available. . . ."[19] In order to respond to this existing and growing demand, the chapter emphasizes that family planning services and supplies will need to be substantially increased now and in the two decades ahead. The chapter also highlights the importance of voluntary choice in family planning, repudiates any form of coercion, and states emphatically that abortion should not be promoted as a means of family planning.[20]

Chapter 8 addresses various health concerns. It challenges nations to make primary health care universally available by the year 2000 and to achieve a life expectancy average of seventy by 2005, and seventy-five by 2015.[21] In the area of child survival and health, the chapter challenges nations to achieve the goals of the 1990 World Summit for Children, which call for a reduction in infant and child mortality by one-third or more by the year 2000. On the matter of women's health and safe motherhood, the chapter calls upon nations to reduce 1990 levels of maternal mortality by 50 percent by the year 2000 and to cut these levels in half again by the year 2015. The chapter also gives special attention to the maternal health concerns surrounding illegal and unsafe abortions. Studies indicate that of the 40 million abortions performed each year, only 26–31 million are performed legally, and approximately 20 million are conducted in unsafe conditions, leading to 67,000 deaths each year.[22] Reiterating the concern that abortion should never be promoted as a means of family planning, the *Programme of Action* makes the following statement about unsafe abortion:

> All governments and relevant intergovernmental organizations are urged to strengthen their commitment to women's health, to deal with the health impact of unsafe abortion as a major public health concern and to reduce the recourse to abortion through expanded and improved family-planning services. Prevention of unwanted pregnancies must always be given the highest priority and every attempt should be made to eliminate the need for abortion. . . . In circumstances where abortion is not against the law, such abortion should be safe. In all cases, women should have access to quality services for the management of complications arising from abortion. Post-abortion counseling, education and family-planning services should be offered promptly, which will also help to avoid repeat abortions.[23]

This paragraph in its draft versions became the center of much controversy in Cairo and threatened to derail the entire conference when the

Holy See together with some other predominantly Roman Catholic nations objected to the wording of the paragraph and the lack of a definition of unsafe abortion. A small but important committee was appointed to redraft the paragraph. In the end, its revisions were minor and essentially involved moving the condemnation of abortion as a means of family planning to the beginning of the paragraph and including a definition of unsafe abortion provided by the World Health Organization (WHO). Note 20 in the *Programme of Action* thus defines unsafe abortion as "a procedure for terminating an unwanted pregnancy either by persons lacking the necessary skills or in an environment lacking the minimal medical standards or both."[24]

Finally, chapter 8 concludes by addressing the various health challenges posed by the HIV/AIDS crisis. The chapter calls all nations to join together to help reduce the spread of the disease, to provide adequate medical care and counseling to those with the disease, to give special attention to the needs of orphans, and to respect and protect the rights of those who are suffering from the disease.

Turning to education, chapter 11 addresses the linkages between population, development, and education. Emphasizing that increases in the education of women and girls contribute to greater empowerment of women, postponement of the age of marriage, and reduction in the size of families, the chapter calls for universal access to primary education for both girls and boys as soon as possible, and no later than 2015.[25] The chapter also calls on nations to close the gender gap in primary and secondary education by 2005.[26] With this summary of the key features of the *Programme of Action*, we can now turn our attention to a moral appraisal of this approach.

Eco-Justice and the *Programme of Action*

Before engaging in a moral assessment of the *Programme of Action* from an eco-justice perspective, it is helpful to summarize briefly the different ways in which moral reflection has been brought to bear on population policies. As we have seen, population policy has gone through some significant changes over the last twenty years. With the development of various ill-conceived and invasive policies in the 1960s and 1970s, much moral reflection during that period and into the 1980s focused on the various types of incentives, disincentives, and forms of coercion that these policies often employed. In evaluating these population policies and methods, ethicists primarily considered their impact on four primary human values: freedom, justice, general welfare, and security or survival.[27]

While responsible moral evaluation of specific population policies in-volved reflection on all four of those primary values, ethicists arranged the values in different orders of priority. For example, some made the values of general welfare and security/survival subordinate to the more fundamental values of freedom and justice. Ethicists such as Ronald Green and Daniel Callahan argued that efforts to maximize freedom and equality are more effective means of securing the common good than coercive means that violate human dignity.[28] Callahan warned that "the end—even security-survival—does not justify the means when the means violate human dignity and logically contradict the end."[29] Others took the opposite approach and emphasized that without a fun-damental measure of general welfare and security it is impossible to experience the values of freedom and justice. Ethicists such as Michael Bayles and James Gustafson argued that it may be necessary and justi-fiable to limit certain individual rights and regulate human fertility out of a concern for the common good of present and future generations.[30] Bayles writes: "Inequality and lack of freedom are acceptable if without them people would lack welfare or security."[31]

Given the many violations of human rights that population policies have perpetrated in the past, and reflecting the new emphasis on the empowerment of women and the need to integrate population policies within the broader task of sustainable development, much contempo-rary moral reflection on population policies, while condemning coer-cion, eschews reflection on incentives and disincentives and prefers to focus on the values of equity, equality, reproductive rights, and bodily integrity.[32] For example, Vandana Shiva and other leading feminists have emphasized that population policies that do not place the well-being of women at their center are hopelessly flawed.[33] A 1992 decla-ration by women's health advocates from Asia, Africa, Latin America, the Caribbean, the United States, and Western Europe stressed that women must be the subjects and not the objects of population policies. The declaration also emphasized the right of women to decide when, whether, and with whom they care to express their sexuality. The dec-laration insisted that these fundamental reproductive and sexual rights should not be subordinated to anyone, including governments.[34]

In my own work on the challenges posed by population growth and overconsumption, I have shared many of the perspectives that I have just described.[35] I have emphasized that any moral response to popu-lation and consumption issues must be grounded in a commitment to social and economic justice. Such a commitment would necessitate a variety of reforms that, in Gandhi's words, would "put first those that society puts last." In this regard, I have argued that the most important area for reform involves empowering women and improving their status in society. Regarding family planning, I have emphasized the need to achieve universal access and voluntary participation. I have also abhorred coercion and have been reluctant to discuss criteria for the evaluation of incentives and disincentives. Finally, I have resisted those who wish to pit the welfare of growing numbers of poor human beings against the welfare of other species. I believe that such discussions wind up "blaming the victims" and do not address the unjust distributions of wealth and power that perpetrate and perpetuate the myriad social and ecological injustices that we now commonly refer to as the eco-crisis.

In what follows, I am not backing away from these views. Instead, I am hoping to expand on them and enrich the quality of moral reflection on population and consumption issues by evaluating the *Programme of Action* on the basis of the concept of eco-justice and its four accompanying norms.

Turning first to the norm of participation, even a cursory reading reveals that this norm is well reflected throughout the *Programme*. The recurring emphases on the empowerment of women, reproductive rights, and the importance of education reveal a deep commitment to the norm of participation. The goal is not only to improve the status of women, but also to empower them. Impressively, the document moves beyond rhetoric toward the substance of those specific measures that will be necessary to realize those goals. This is most clearly seen through its emphasis on universal access to reproductive health care, elimination of various obstacles to the full participation of women in society, and improvements in education at the primary and secondary levels.

At the heart of this new approach to population and development issues rests the goal of promoting, strengthening, and maintaining the moral agency of fully one half of the human species. In its championing of reproductive rights, however, the document attempts to avoid atomistic and individualistic misconceptions by emphasizing the need to balance these rights with the responsibilities one owes to living and future children as well as the broader community.[36] With its opposition to coercion and its reluctance to embrace various incentive or disincentive schemes, the *Programme of Action* distances itself from controversial policies like those adopted in China over the last twenty years to regulate human fertility. Given the kinds of human rights abuses that have taken place in China, this approach is commendable.

The reality remains, however, that some nations with high rates of fertility, population growth, and demographic momentum may not see these rates drop fast enough through the means proposed in the *Programme of Action* to avoid social, political, and ecological collapse. The British scholar Maurice King has argued that faced with the doubling of their populations in approximately twenty years, lacking the resources to either grow or trade for food, and faced with the reality that mass internal or international migration is not an option, nations like Rwanda, Malawi, Haiti, and Bangladesh face "demographic entrapment." In order to avert social, political, and ecological disaster, King argues that these nations should implement a one-child policy or a substantial mix of incentives and disincentives to reduce fertility and slow demographic momentum.[37] King's claim is that in such severe cases, the restriction of reproductive rights and freedoms will do less harm than the consequences of unregulated fertility and the perpetuation of the status quo.

King's arguments raise the issue of the relationship of individual rights to the powers of governments to limit these rights out of a need to protect the common good. While the power of the state to limit the free exercise of rights is clearly established and practiced throughout the world, I believe it is less clear that the need exists for nations to limit individual reproductive freedoms. Few nations are in such dire straits as Rwanda or Haiti, and thus a one-child policy or aggressive incentives and disincentives would be unjustified in most nations in the world. In order for such a policy even to be considered, at least the following criteria need to be satisfied. First, it must be demonstrated that all measures in the *Programme of Action* have been tried and found insufficient. In addition, it must be demonstrated that efforts to permit international migration and achieve food security have been attempted through the international community and have failed. Any decision to implement such drastic measures as a one-child policy or a similar set of incentives and disincentives to reduce fertility must be the result of a democratic process that spreads the burdens and responsibilities equally across the entire population. Measures must be taken to prohibit sex selection and discrimination against girl children. Finally, reversible means of contraception must be used, and forced contraception, sterilization, and abortion must be condemned. Rather than controlling fertility, population policies in the twenty-first century need to enhance reproductive freedoms and empower women and men to make responsible decisions regarding their sexual and reproductive lives. This issue will be revisited in the discussion regarding the norm of sustainability.

The norm of sufficiency is also well reflected in the *Programme of Action*. The claim that development is a fundamental human right and the document's related emphases on poverty eradication, gender equity, education, and universal access to primary and reproductive health care all demonstrate that improving the quality of life through the provision of basic human needs is the primary goal of all efforts at sustainable development. The primary issue concerning the norm of sufficiency is not whether the *Programme of Action* embraces the goals that reflect this norm but whether the means to achieve those goals will be sufficient.

Specifically, the major question is whether the funding commitments made in Cairo will be sufficient to accomplish the objectives defined in the *Programme of Action*. Chapter 13 lays out the cost figures (in U.S. dollars) and fiscal obligations:

> It has been estimated that, in the developing countries and countries with economies in transition, the implementation of programmes in the area of reproductive health, including those related to family planning, maternal health and the prevention of sexually transmitted diseases, as well as other basic actions for collecting and analyzing population data, will cost: $17 billion in 2000, $18.5 billion in 2010 and $21.7 billion in 2015.
>
> It is tentatively estimated that up to two thirds of the costs will continue to be met by the countries themselves and in the order of one third from external sources. However, the least developed countries and other low-income developing countries will require a greater share of external resources on a concessional and grant basis.[38]

The first problem with these cost figures is that while they may be sufficient to achieve universal access to family planning, provide basic reproductive care, and prevent the spread of HIV/AIDS and other sexually transmitted diseases, they do not address many of the other significant objectives outlined in the *Programme of Action*. For example, the document acknowledges that additional funds will be required to eradicate poverty, improve the status of women, and accelerate sustainable development programs. More specifically, additional funds will be required to improve primary health care systems, achieve the child mortality and universal basic education goals, and cope with the social and economic costs produced by the AIDS pandemic.[39] It is not clear how much it will cost to achieve those goals, but it is clear that it will be far in excess of the $17 billion slated for the year 2000.

This fact substantially tempers the significance of the consensus achieved in Cairo. On the one hand, the international community has agreed that commitments to the empowerment of women and reproductive health should serve as the primary foundations for a population policy and that such policies must be framed within the broader context of the quest for sustainable development, which seeks to eradicate poverty and improve the quality of lives of all. On the other hand, by fully funding only the goal of universal access to family planning, the international community has adopted a funding commitment that is far less radical than the vision it proposed in the *Programme of Action*. This gap between rhetoric and reality presents the largest obstacle to implementation of the Cairo consensus and raises substantial questions regarding the sufficiency of the approach.

The second major problem with the funding estimates concerns the likelihood that the international community will not live up to its financial commitments in the family planning, reproductive, and sexual health areas. The *Programme of Action* calls for funding in these areas to increase from estimated 1995 levels of $4.5 billion to the $17 billion goal projected for the year 2000. In September of 1994, it appeared that funding in this area would increase. Just prior to and during the conference, several developed nations announced their intention to increase significantly the amount of funds devoted to these goals. The United States doubled its 1992 levels of population assistance to almost $600 million for 1995 and announced its intention to increase that commitment to $1.2 billion by the year 2000. Japan pledged $3 billion over seven years (an average of approximately $430 million per year), and Germany announced a similar commitment of $2 billion over the same period of time (an average of $285 million per year).[40] Unfortunately, these commitments fall well short of the nearly sixfold increase and $5.7 billion that developed nations have agreed to provide by the year 2000.[41] In addition, at least in the United States, increased levels of population and development assistance are in substantial jeopardy as Congress debates various deficit reduction measures that would substantially curtail U.S. foreign aid contributions.[42] Some members of Congress have even called for the United States to withdraw all of its support from the United Nations Population Fund (UNFP) because of its relationship to China's controversial population policies. Given UNFP's major international leadership role, that would be an unmitigated disaster.

These are sobering realities. If developed nations cannot hold up their end of the bargain, how fair is it to expect poorer, developing nations to raise the remainder of the $17 billion goal? Given the social, economic, and ecological challenges that rapid population growth poses over the next fifty years, an international investment of $17 billion in universal access to family planning and reproductive care is a sheer bargain, especially when compared to annual military expenditures that approach $800 billion per year. One thing is clear: In order for the Cairo consensus to work, substantial effort must now be devoted to generating sufficient financial resources to fund all of the objectives contained in the *Programme of Action*. With current political realities, that will be no small task.

Turning to the norm of sustainability, two major questions emerge. First, how well does the Cairo consensus address the dual set of challenges posed by population growth and overconsumption? Second, will the measures proposed in Cairo be sufficient to stabilize the human population at sustainable levels of production, consumption, and reproduction?

In response to the first question, one of the clear weaknesses of the Cairo consensus is that it is far more specific regarding measures that need to be taken to address population growth than it is on measures necessary to address unsustainable levels of production and consumption. It also gives little, if any, specific attention to issues that are particularly important among environmentalists, such as the preservation of wilderness areas, the protection of biodiversity, air and water pollution, soil erosion, and land degradation.

While these and other issues were the primary focus of the Earth Summit, the *Programme of Action* does not provide much new or additional reflection on these issues. It does, however, emphasize the need for nations in the north to reduce their rates of consumption and production and also to help establish macroeconomic conditions that will foster sustained economic growth within the context of sustainable development. In contrast, however, to other sections of the *Programme of Action* that offer specific goals in areas like family planning and primary education, the document emphasizes only the need to "modify unsustainable consumption and production patterns through economic, legislative and administrative measures."[43]

In the end, it is fair to say that the document gives only lip service to consumption issues. While this is an improvement over previous documents, it also reflects the fact that powerful nations like the United States and many in the European Economic Union are not eager to discuss and address these issues. This inability to address economic inequities and ecological costs was not unique, however, to the ICPD. It is a problem at all UN conferences, and it would be unrealistic to expect a substantial departure from that norm in Cairo. Nevertheless, somehow consumption issues and economic disparities must be addressed. To do that will require challenging the god of economic growth and the vested powers that have much to lose from any radical restructuring of economic activity and redistribution of wealth and capital.

Turning to the question of whether the measures adopted in the *Programme of Action* will stabilize global population at a size that is ecologically sustainable and just, the document does not cite a specific figure or goal, but it does make the following claim:

> Implementation of the goals and objectives contained in the present 20-year *Programme of Action* . . . would result in world population growth during this period and beyond at levels below the United Nations medium projection.[44]

Earlier in 1994, however, the United Nations Population Fund released a projection that claimed that the kinds of measures being considered for the *Programme of Action* would result in a low-growth path that would stabilize world population at 7.8 billion in 2050 rather than the 10 billion estimated in the UN's medium variant projection.[45] Given the altered age structure and level of demographic momentum in many nations in the developing world, stabilization of global population at 7.8 billion through voluntary means would be a significant achievement. While it is clear that global food supplies and ecological problems pose significant challenges ahead, it will be better to face these challenges with a stabilized population of approximately 8 billion than with a population that continues to grow during the next century and that may stabilize closer to 12 billion in 2100, if then.

The key factor, however, is whether the world will really embrace the approach to population and development that its leaders adopted in Cairo. Beyond the concerns regarding funding rests another critical concern regarding political will. The very first sentence of chapter 2, which frames the principles that undergird the *Programme of Action*, makes the following proviso:

> The implementation of the recommendations contained in the *Programme of Action* is the sovereign right of each country, consistent with national laws and development priorities, with full respect for the various religious and ethical values and cultural backgrounds of its people, and in conformity with universally recognized international human rights.[46]

Some have pointed to this stipulation as the major loophole in the document, which allows nations to pick and choose among the important objectives contained in the *Programme of Action*. Certainly, conservative Roman Catholic and Islamic countries made it clear that there were portions of the document that they could not support on the basis of their religious and cultural beliefs. Does this mean that the *Programme of Action* is fatally flawed and will not work? I do not think so. It will reduce the effectiveness of the plan in some nations, but it will not substantially thwart the overall gains that the approach offers. Despite the predominantly Roman Catholic composition of Latin America, population growth and fertility rates have declined as access to modern means of family planning has increased. While many Islamic nations have some of the highest fertility rates in the world, it is also clear that many of them have found ways to embrace family planning methods while maintaining Islamic law and culture. Finally, while 10 percent of the 179 nations and six observers offered reservations on various portions of the *Programme of Action,* it is important to emphasize that 90 percent of the delegations representing 95 percent of the world's population fully endorsed the program.

It is one thing, however, to conclude that the Cairo consensus can work, but it is another thing to decide whether it will work fast enough. Given the current levels of poverty and malnutrition, worrisome projections regarding food supply, and alarming rates of loss in biodiversity, topsoil, and other renewable and nonrenewable resources, some suggest that the *Programme of Action* is necessary but not sufficient to stabilize global population as soon as possible and at as low a level as possible. Toward this end, critics contend that more "direct" measures need to complement the "indirect" measures reflected in the Cairo consensus.[47] These direct measures involve various incentives and disincentives aimed at individuals and communities to impact decisions and behavior regarding fertility. Various incentives might include free sex education and family planning services, free sterilization or contraceptive implantation, various cash or material incentives, tax breaks, or housing privileges. Various disincentives might include fewer tax deductions, monetary penalties, or the revocation of various social privileges for fertility above the stated goal.[48]

The *Programme of Action* takes the following position on incentives and disincentives:

> The principle of informed free choice is essential to the long-term success of family-planning programmes. Any form of coercion has no part to play. In every society there are many social and economic incentives and disincentives that affect individual decisions about child-bearing and family size. Over the past century, many governments have experimented with such schemes, including specific incentives and disincentives, in order to lower or raise fertility. Most such schemes have had only marginal impact on fertility and in some cases have been counterproductive. Governmental goals for family planning should be defined in terms of unmet needs for information and services. Demographic goals, while legitimately

the subject of government development strategies, should not be imposed on family-planning providers in the form of targets or quotas for the recruitment of clients.[49]

Governments are encouraged to focus most of their efforts towards meeting their population and development objectives through education and voluntary measures rather than schemes involving incentives and disincentives.[50]

Not surprisingly, most recent feminist reflection on population policy has been quite critical of incentive and disincentive schemes.[51] I share this wary approach to "direct" measures and believe that the objectives framed in the *Programme of Action* need to be attempted before such schemes should be considered. In addition, I believe that any set of incentives or disincentives needs to be designed, implemented, and evaluated by those most directly affected, who—in almost all cases—will be women. On the other hand, I believe that Sonia Correa and Rosalind Petchesky are right when they suggest that it is necessary to "distinguish . . . *supportive* or *empowering* conditions from *coercive* incentives or disincentives. . . ."[52] Some incentives actually increase the moral agency of women rather than undermine it. The key is that women must have the power to decide. Given the past abuses of various incentive and disincentive schemes, it is certainly not wise or necessary to pursue these schemes in any aggressive fashion. Moreover, appeals to such measures on the grounds of ecological well-being tend to miss entirely the devastating impact of small and relatively stable populations in the developed world that are wreaking havoc on the earth through their unsustainable patterns of production and consumption. It seems to me that proponents of "direct" measures should focus their attentions there.

This brings us finally to the norm of solidarity. The *Programme of Action* emphasizes that "international cooperation and universal solidarity" will be critical to the implementation of the Cairo consensus.[53] The document repeatedly calls upon all nations to listen to the cries of the poor and the marginalized and to respond by attempting to eradicate poverty, empower women, and protect children, the elderly, and the sick and disabled. In addition, through its emphasis on sustainable development, the *Programme of Action* calls attention to the welfare of the planet and the need to live in solidarity with all of those living on the earth. While the document clearly has an anthropocentric focus, it is infused with the belief that greater measures of social justice will lead to greater measures of ecological integrity.

This rhetoric of solidarity, however, is undermined in two very important ways. Given the great disparities of wealth and power in the world, it is unjust that developing nations are saddled with two-thirds of the financial burden of implementing the *Programme of Action*. In relationship to this issue, the proposal for nations in both developing and developed nations to spend 20 percent of their budget in the social sector was debated but not affirmed in Cairo. It should be noted, however, that this "20/20 Proposal" was accepted at the World Summit on Social Development, which was held in Copenhagen in March 1995. This is an improvement and an expression of solidarity, but there is good reason to doubt that either developed or developing countries will live up to this commitment. It is difficult to see solidarity reflected in U.S. budgetary priorities, which over the last decade have limited U.S. foreign assistance to less than 1 percent of the federal budget, with nearly half of those funds taking the form of national security assistance to Israel and Egypt. In 1996, the U.S. Congress reduced the humanitarian and development assistance portions of U.S. foreign aid in 1996 by approximately 30 percent.

Another failure to match substance with the rhetoric of solidarity is in the insufficient attention the *Programme of Action* gives to the challenges posed by overconsumption and unsustainable economic activity in developed nations. This concern was raised throughout the preparatory process and in Cairo by official delegates from developing nations and by representatives of various nongovernmental organizations from around the world. Some argued that these inadequacies are inevitable so long as existing capitalist structures, powers, and policies are retained as the primary framework for international economic relations. While I do not believe that most in the international community are prepared to shift from capitalist to socialist conceptions of economic activity, I do believe that there is a groundswell among various groups and sectors in nations like the United States that recognize the need to redefine economic development in terms of sustainable development. Within this new view of economic activity, economic growth and welfare would be measured in more qualitative terms than in the standard quantitative measurements of GNP or GDP. For all of its faults, the President's Council on Sustainable Development (PCSD) has attempted to outline such a sustainable development path for the United States. Established by President Clinton on Earth Day in 1993, the work of the council has been structured around the three goals of preserving ecological integrity, increasing social equity, and ensuring economic prosperity. Having served on the Population and Consumption Task Force of the council, I must say that I believe the council has done a better job of addressing its economic and ecological objectives than of addressing its goal to increase social equity. This dimension of sustainable development either was absent in much conversation or took a back seat to economic and environmental concerns. This is also what took place in Cairo. While social equity was addressed through the *Programme of Action*'s emphasis on gender equity, health care, poverty, and education, little attention was given to the disproportionate ecological impact of the highly consumptive lifestyles of wealthy citizens living in nations like the United States. This is not surprising, but from an eco-justice perspective it is a major flaw. Drawing upon the norm of solidarity, religious communities, and Christians in particular, must continue to raise this concern in national and international policy discussions that impact the poor.

Conclusion

This brings me to the conclusion of this eco-justice critique of the *Programme of Action* adopted at the 1994 United Nations International Conference on Population and Development. Clearly, this has been a mixed moral assessment. While there is much to be commended in the proposals adopted in Cairo, there are also significant faults and flaws that need to be addressed. It is not necessary to reiterate all of these views here. Nevertheless, I do believe that the Cairo consensus really does represent a watershed shift in comparison to the policies that preceded it. If the goals of the *Programme of Action* are achieved over the course of the next two decades, a greater measure of social justice and ecological integrity should prevail. It remains to be seen, however, whether the fourth and fifth decennial United Nations conferences on population and development in 2004 and 2014 will be occasions for celebration or mourning over lost opportunities.

As I look forward to the twenty-first century, the greatest source of hope for me is found not in the *Programme of Action* approved in Cairo, however, but rather in having witnessed the exuberant and intelligent contributions of women and men around the world as they worked together to address these issues and forge a new consensus. While much more reflection needs to be given to consumption issues, it is clear that a new "sensibility" is emerging in the international community and within the churches that recognizes the need to pursue the twin goals of ecological integrity and social justice. This vision is finding expression in diverse efforts to promote sustainable development, the eradication of poverty, the empowerment of women, and improvement in the quality of life for all who live on the earth. Those who champion the concept of eco-justice should draw comfort from this reality and lend their efforts, in both constructive and critical ways, to further developing this emerging sensibility and making it manifest in concrete ways.

Notes

1. I had the privilege of attending the conference as a member of a nongovernmental delegation sponsored by the Religious Consultation on Population, Reproductive Health, and Ethics.

2. See William E. Gibson, "Eco-Justice: New Perspectives for a Time of Turning," in Dieter Hessel, ed., *For Creation's Sake: Preaching, Ecology, and Justice* (Philadelphia: Geneva Press, 1985), pp. 15–31.

3. The fifth assembly of the World Council of Churches in 1975 emphasized the need to create a "just, participatory, and sustainable society." A follow-up conference in 1979 entitled "Faith, Science and the Future" gave explicit attention to the norms of sustainability, sufficiency, participation, and solidarity. In 1983, the sixth assembly of the WCC challenged all of its member communions to strive for the integration of "justice, peace, and the integrity of creation." This emphasis continued in 1990 with the theme of the seventh assembly, "Come Holy Spirit—Renew Your Whole Creation."

4. See Presbyterian Eco-Justice Task Force, *Keeping and Healing the Creation* (Louisville, Ky.: Committee on Social Witness Policy, Presbyterian Church (USA), 1989); and Evangelical Lutheran Church in America, *Caring for the Creation: Vision, Hope, and Justice* (Chicago: Division for Church in Society, 1993). The term is also being used by various groups including the Eco-Justice Working Group of the National Council of Churches and the Eco-Justice Project and Network of the Center for Religion, Ethics, and Social Policy at Cornell University.

5. For a biblical and theological foundation for these norms, see James B. Martin-Schramm, "Toward an Ethic of EcoJustice," in Paul Jersild, Dale Johnson, Patricia Jung, and Shannon Jung, eds., *Moral Issues & Christian Response*, 6th edition (Fort Worth: Harcourt Brace, 1998), pp. 208–13.

6. Eighteen predominantly Roman Catholic or conservative Islamic nations expressed final reservations on specific chapters or paragraphs of the *Programme of Action*. Nations such as El Salvador, Argentina, and Ecuador emphasized that in no way did they wish to commend abortion as a means of birth control. Nations such as Yemen, Iran, and the United Arab Emirates emphasized that they could not embrace those portions of the *Programme of Action* that in their view violated Islamic law, giving particular attention to calls for changes in inheritance laws. Both groups emphasized marriage as the sole appropriate context for sexual activity and expressed concern about language that emphasized the reproductive rights of not only "couples" but also of "individuals." In contrast to the plans developed at predecessor conferences, no nation rejected the entire document. The Holy See, however, did join the consensus on only six of the sixteen chapters. Nevertheless, the 179 delegations fully endorsing the *Programme*

of Action represented 90 percent of the delegations and 95 percent of world population.

7. There are now several excellent assessments of the ICPD; see in particular: Lincoln C. Chen, Winifred M. Fitzgerald, and Lisa Bates, "Women, Politics, and Global Management," *Environment* 37, no. 1 (January–February 1995): 4–9, 31–33; Gita Sen, "The World Programme of Action: A New Paradigm for Population Policy," *Environment* 37, no. 1 (January–February 1995): 10–15, 34–37; and Adrienne Germain and Rachel Kyte, *The Cairo Consensus: The Right Agenda for the Right Time* (New York: International Women's Health Coalition, 1995). I have also offered a brief assessment entitled "Cairo Beyond Abortion" in *Sojourners* 23, no. 9 (November 1994): 10–11.

8. Stanley P. Johnson, *World Population and the United Nations: Challenge and Response,* (New York: Cambridge University Press, 1987), p. 65. See also Nafis Sadik, *Population Policies and Programmes: Lessons Learned from Two Decades of Experience* (New York: UNFPA, New York University Press, 1991); and also by Nafis Sadik, "The Role of the United Nations: From Conflict to Consensus," in Godfrey Roberts, ed., *Population Policy: Contemporary Issues* (New York: Praeger Press, 1990).

9. Lincoln C. Chen, Winifred M. Fitzgerald, and Lisa Bates, "Women, Politics, and Global Management," pp. 5–6.

10. Excerpt of speech by U.S. representative James L. Buckley to the 1984 conference, quoted in Stanley P. Johnson, *World Population and the United Nations,* pp. 255–56.

11. Peter J. Donaldson, *Nature Against Us: The United States and the World Population Crisis* (Chapel Hill: University of North Carolina Press, 1990), p. 131. For a detailed, albeit vested, defense of UNFPA's role in China's family planning program, see Stanley P. Johnson, "China, the United States, and the United Nations," *World Population and the United Nations,* pp. 288–312.

12. For an excellent summary, see Gita Sen, "The World Programme of Action: A New Paradigm for Population Policy."

13. United Nations of the International Conference on Population and Development, *Programme of Action* (New York: United Nations Population Fund, September 1994), p. 21, par. 3.16.

14. Ibid., p. 25, par. 4.1.

15. Ibid., p. 25, par. 4.2.

16. Ibid., p. 43, par. 7.2.

17. Ibid., p. 43, par. 7.3.

18. Ibid., p. 44, par. 7.6.

19. Ibid., p. 46, par. 7.13, emphasis added.

20. Ibid., pp. 47–48, pars. 7.15, 7.22, 7.24. Contrary to criticism of the *Programme of Action,* this opposition to abortion as a means of family planning was clearly stated in the draft version of the document that was developed before the meeting of the ICPD in Cairo.

21. Ibid., p. 56, par. 8.5. These goals are reduced to sixty-five and seventy years respectively for nations that currently have high mortality rates.

22. Nafis Sadik, *The State of World Population 1995* (New York: United Nations Population Fund, 1995), p. 6.

23. United Nations, *Programme of Action*, pp. 61–62, par. 8.25.

24. Ibid., p. 118. This move by the Holy See triggered several harsh responses, and many openly questioned the Vatican's privileged "permanent observer" status within the United Nations. More positively, the Vatican position generated substantial reflection and discussion on the views of other world religions regarding the substance and issues posed by the draft *Programme of Action*. This was especially true at the NGO Forum, which was being held across the street from the official UN conference site. The NGO Forum was attended by four thousand delegates from 133 nations, and here, various groups sponsored panel presentations. Religious groups offered Buddhist, Jewish, Muslim, and other Christian perspectives on family planning, abortion, gender equality, and other related issues. Representatives from the World Council of Churches, the National Council of Churches, and the Religious Consultation on Population, Reproductive Health, and Ethics held a joint news conference, which was packed with reporters eager to hear alternative religious perspectives on the ICPD.

25. Ibid., p. 80, par. 11.6.

26. Ibid., p. 80, par. 11.8.

27. For a broader discussion of the meaning of these key ethical terms, see Robert Veatch, ed., *Population Policy and Ethics: The American Experience* (New York: Irvington Publishers, 1977), pp. 17–52; and Donald Warwick, "The Ethics of Population Control," in Godfrey Roberts, ed., *Population Policy: Contemporary Issues* (New York: Praeger Press, 1991), pp. 22–23. For an excellent example of Christian moral evaluation of population policies using these four primary values, see "United States Population Policy and the Church," a statement adopted in 1972 by the 184th General Assembly of the United Presbyterian Church in the USA, in *Minutes of the General Assembly* (New York: UPCUSA, 1972), pp. 12–59.

28. See Ronald Michael Green, *Population Growth and Justice: An Examination of Moral Issues Raised by Population Growth* (Missoula, Mont.: Published by Scholars Press for Harvard Theological Review, 1976); and Daniel Callahan, "Ethics and Population Limitation," in Michael D. Bayles, ed., *Ethics and Population* (Cambridge, Mass.: Schenkman Publishing, 1976).

29. Callahan, "Ethics and Population Limitation," pp. 34–35.

30. See Michael D. Bayles, *Morality and Population Policy* (Birmingham: University of Alabama Press, 1980); and James M. Gustafson, "Population and Nutrition," *Ethics from a Theocentric Perspective*, vol. 2 (Chicago: University of Chicago Press, 1984).

31. Bayles, *Morality and Population Policy*, p. 35.

32. For an excellent discussion of this transition and an outline of the new directions moral philosophers are taking on population policy, see Sissela

Bok, "Population and Ethics: Expanding the Moral Space," in Gita Sen, Adrienne Germain, and Lincoln C. Chen, eds., *Population Policies Reconsidered: Health, Empowerment, and Rights* (Boston: Harvard Center for Population and Development Studies, 1994), pp. 15–26. On the theological ethics side, Susan Power Bratton's *Six Billion & More: Human Population Regulation and Christian Ethics* represents an interesting example of moral reflection that was conducted during this period of transition. Bratton's work reflects the emphasis on social and economic justice that began to emerge during the 1980s on these issues, but she also includes substantial discussion of incentives and disincentives that are less favored today.

33. Maria Mies and Vandana Shiva, *Ecofeminism* (London: Zed Books, 1993), pp. 277–96.

34. International Women's Health Coalition Secretariat, "Women's Declaration on Population Policies," in Sen, Germain, and Chen, eds., *Population Policies Reconsidered: Health, Empowerment, and Rights*, pp. 31–34. See also Sonia Correa and Rosalind Petchesky, "Reproductive and Sexual Rights: A Feminist Perspective," in that volume, pp. 107–23. A similar declaration was drafted in 1994 by an international group of ethicists prior to the ICPD. In their declaration, those ethicists identified five ethical propositions that should govern moral reflection on population policies. These policies should (1) promote the concepts of reproductive health and reproductive rights; (2) advocate the equitable allocation of benefits and responsibilities related to reproductive decisions; (3) emphasize respect for persons and the principle of autonomy; (4) achieve more desirable consequences over those that achieve less desirable ones; and (5) integrate population policies with other attempts to improve the quality of life for all people. See United Nations Population Fund, "Declaration of Ethical Propositions," Stephen L. Isaacs, ed., *Ethics, Population and Reproductive Health Roundtable* (New York: Development Law & Policy Program, Columbia University, 1994), pp. 1–17.

35. James B. Martin-Schramm, "Population Growth and Consumption Issues: The State of the Debate in the Field of Christian Ethics," in Dieter Hessel, ed., *Theology for Earth Community: A Field Guide* (Maryknoll, N.Y.: Orbis, 1996); "Population Policies and Christian Ethics," in Laurie Mazur, ed., *Beyond the Numbers: A Reader on Population and Consumption Issues* (Washington, D.C.: Island Press, 1994), pp. 310–17; "Population Growth and Justice," panel remarks presented at the third preparatory meeting for the International Conference on Population and Development, held at the United Nations, May 19, 1993, in *Religious and Ethical Perspectives on Population Issues* (Washington, D.C.: Religious Consultation on Population, Reproductive Health, and Ethics, 1993), pp. 12–16; "Justice as the Goal," *The Egg: An Ecojustice Quarterly* 12, no. 4 (Fall 1992): 11; and "Population Growth, Poverty, and Environmental Degradation," *Theology and Public Policy* 4, no. 1 (Summer 1992): 26–38.

36. United Nations, *Programme of Action*, p. 43, par. 7.3

37. Maurice King, "Lessons from Rwanda: The Case for a One-Child World," unpublished paper presented at the NGO Forum during the United Nations International Conference on Population and Development in Cairo, Egypt, September 5–13, 1994, pp. 1–4.

38. United Nations, *Programme of Action*, pp. 97–98, pars. 13.15 and 13.16. These funding goals (adopted by the 179 nations that approved the *Programme of Action*) are broken down in the following way:

	2000	2005	2010	2015
	U.S. Dollars (in billions)			
Family planning	10.2	11.5	12.6	13.8
Reproductive health	5.0	5.4	5.7	6.1
STD and HIV/AIDS	1.3	1.4	1.5	1.5
Research and data	.5	.2	.7	.3

39. Ibid., pp. 98–99, pars. 13.17–13.20.

40. Susan Kalish, "Cairo Built Momentum for Change, Say Advocates," *Global Stewardship Network* (October–November 1994): 2.

41. See Lincoln C. Chen, Winifred M. Fitzgerald, and Lisa Bates, "Women, Politics, and Global Management," p. 32; and Lori S. Ashford, "New Perspectives on Population: Lessons from Cairo," *Population Bulletin* 50, no. 1 (March 1995): 36.

42. Susan Kalish, "Will the U.S. Renege on Cairo Funding Commitments?" *Global Stewardship Network* (April–May 1995): pp. 1–2.

43. Ibid., p. 24, par. 3.29(d).

44. United Nations, *Programme of Action*, pp. 9–10, par. 1.4.

45. Paul Lewis, "U.N. Conference to Discuss Population Plan," *New York Times,* April 3, 1994. A study produced by John Bongaarts, director of the research division at the Population Council, is even more optimistic. He projects that vigorous implementation of the objectives reflected in the *Programme of Action* could reduce global population to a level slightly *less* than existing levels. The problem with his projections, however, is that he assumed that universal access to family planning and education would be achieved in 1995 rather than 2015, which was obviously an unrealistic assumption. See John Bongaarts, "Population Policy Options in the Developing World," *Science* 263 (February 11, 1994): 771–76.

46. United Nations, *Programme of Action*, p. 14.

47. See Virginia D. Abernathy, *Population Politics: The Choices That Shape Our Future* (New York: Plenum Press, 1993), pp. 73–84; William G. Hollingsworth, *Ending the Explosion: Population Policies and Ethics for a Humane Future,* unpublished manuscript (1995); Maurice King, "Lessons from Rwanda: The Case for a One-Child World," pp. 1–4; and Garrett Hardin, *Living Within Limits: Ecology, Economics, and Population Taboos* (New York: Oxford University Press, 1993), pp. 272–73.

48. For a more specific discussion of the types of incentives and disincentives offered by nations, see Nafis Sadik, *Population Policies and Programmes:*

Lessons Learned from Two Decades of Experience, pp. 120–23; see also Robert Veatch, "An Ethical Analysis of Population Policy Proposals," *Population Policy and Ethics: The American Experience,* pp. 445–75.

49. United Nations, *Programme of Action,* p. 46, par. 7.12.

50. Ibid., p. 48, par. 7.22.

51. See Sissela Bok, "Population and Ethics: Expanding the Moral Space," p. 24; International Women's Health Coalition Secretariat, "Women's Declaration on Population Policies," p. 33; and United Nations Population Fund, "Declaration of Ethical Propositions," p. 4.

52. Sonia Correa and Rosalind Petchesky, "Reproductive and Sexual Rights: A Feminist Perspective," p. 116. Emphasis added.

53. United Nations, *Programme of Action,* p. 14.

The Transition to a Transition

Neva R. Goodwin*

Suppose we agreed that it would be desirable—for environmental, or cultural, or spiritual reasons—to change the direction of our society away from a culture of consumerism. Such a transition would require a massive change in many major aspects of our lives. We are so far from making such a change that it is impossible to imagine it as a whole; we can imagine only pieces of it—the pieces that would have to be put in place if we were to make the transition to that transition.

Stretching our imaginations as far as possible, let's suppose that everyone in the United States decided that, in principle, it is a good idea to move toward such a change. What might be the difficulties in the way of acting on this decision? To answer this question, we need to understand our socioeconomy as a coherent, internally consistent whole. It is held together by a core set of interlocking elements that nest in a configuration from which it is hard to move any one without breaking others or disordering the whole pattern.

I will begin by briefly listing these elements and will then describe some of their interlinkages before I turn to the question of how they might be disengaged.

They include, on a broad societal scale:

- the historical fact of how industrialization and economic development have occurred in the last two centuries, strictly depending on

---

*Neva R. Goodwin is codirector of the Global Development and Environment Institute at Tufts University.

the growth of labor productivity, so that over time each employed worker can produce a higher value of output;
- the profit motive as a driver of business;
- the fact that income is so often and so tightly tied to a job as defined in the modern world;
- advertising, as an expression of the power of the corporate interests, and as a strong shaper of arts and media;
- the way (in the United States) that corporate ownership and corporate responsibilities and obligations have been defined, overwhelmingly in relation to stockholder returns;
- political power that is based broadly on perceptions about what it takes to provide jobs and narrowly on the interests of the very large corporations; and
- the way GDP is defined.

On the more individual scale, the list continues with a number of not terribly appealing, but extremely potent, human values that have risen to the fore in modern societies. These include:

- the way personal success has been defined and related to the current concept of corporate success;
- social emulation, or keeping up with the Joneses;
- competitiveness;
- and, as the linchpin, holding all of these elements together: the fact that consumerism has become one of the major values, or value-systems, of our society.

It may not be immediately obvious why and how all of these elements are mutually reinforcing. I will trace through some of the interlinkages, not covering every one of the items I have just listed, but trying to suggest their importance and their impacts.

The Historical Growth of Labor Productivity

The central factor that made the modern consumer society both possible and necessary was the success of the industrial revolution, particularly in raising labor productivity. That may sound dry and economistic. In all sorts of ways it was an extraordinarily dramatic and important event.

The industrial revolution—starting in England over two hundred years ago—incorporated innovations in technology and management that made it possible for the average worker to produce more output each year than she or he had produced the year before. We have become used to the idea that productivity should rise every year, and while we regard with some awe the approximately 10 percent annual rise of productivity in China, when our own growth slows to 2 percent, which we think is pretty disappointing. In fact, a productivity gain of 2 percent a year means that if the number of workers stays the same, in thirty-five years their output will have doubled.

During this century, the output of our economy has doubled many times over. Our population has also doubled several times, but the growth in productivity—and therefore in output—has gone so much faster that output *per person* has kept doubling. This is, essentially, the explanation for the rising standard of living that was so outstanding from the early part of this century up until the 1980s.

One way of gaining a deeper understanding of a historical phenomenon is to ask about it, "Was this inevitable? Could events have played out some different way?" I have emphasized the increase in output that can be expected from one generation of workers. Let us call that the "growth dividend," and let us ask ourselves: What are the different ways that the growth dividend could have been—or could still—be used?

At the beginning of the industrial revolution, the answer to this question was not at all a foregone conclusion. In fact, it might have seemed most likely that societies would continue to use any surplus they could generate in the way they almost always had; that is, by the inegalitarian solution of producing more and more luxuries for a very small elite—maybe 1 percent of the population—while the masses would continue to exist at a subsistence level.

It is not possible, here, to go into the fascinating question of why that is not what happened in the eighteenth century in England, Western Europe, and North America. But we all know what actually did happen: The growth dividend was, instead, gradually turned toward mass consumption.[1]

Before the seventeenth century, a common person would likely own only one, or if he or she was reasonably well off, two suits of clothes during his or her adult lifetime. The number of items owned by a household that was not in the elite could probably be counted on the fingers of both hands. That is hard for us to imagine. The norm was what we would now regard as abject poverty. It is not a norm to which I have any interest in returning. However, the choice of mass consumption has had some side effects that are not altogether desirable.

As we examine these undesirable side effects, we should keep in mind (lest we become too simplistic in our thinking) that industrialized societies still have the option of depending on ever greater luxury for the elite, rather than mass consumption, to soak up our still growing output. Some societies (e.g., in Latin America and India) have long taken the inegalitarian alternative quite for granted. Others, in North America and western Europe, seem to be flirting with the idea of going backward toward the inegalitarian solution. That, too, is a highly unappealing alternative. If the problems of mass consumption, to which I will refer below, are the Scylla away from which we should steer, the inegalitarian solution to the productivity of industrialization is the Charybdis. The path between them is narrow and often hard to discern.

Profits, Jobs, and Advertising: The Culture of Consumerism

Because we took the path that we did—choosing mass consumption as the way to use the growth dividend—our economy became hooked on mass consumption. The logic is simple. As we became more and more productive, more and more things were produced. What is produced *must* be sold—that is the essential requirement. As people's basic needs were increasingly met, there was a concern—especially during the first two decades of this century—that we might be approaching a saturation point. What would then happen to the system that was geared to turning out more and more things? That concern was the basis for the employment of ever more of our society's talent and energies in finding ways to make people want things that they had not thought of wanting. This is not all bad. I did not know I wanted a microwave oven, or a word processor, or modern, easy-to-use cross-country skis. I am delighted that so much ingenuity has gone into inventing and improving these products. A number of large problems remain, however.

One is that within our present system, it appears that production of the things that really enhance life cannot employ everybody—or even half of the people who want employment. Part, though not all, of the reason for this goes back to that remarkable two-century process of rapidly increasing labor productivity. One person can now produce what hundreds or even thousands of people would have been required to produce before industrialization.

To be sure, there are still some very important tasks that are undervalued in our socioeconomic system, in the sense that the wages they command are much lower than would be commensurate with their contribution to human well-being. Low or zero wages—sometimes even financial penalties[2]—are attached to the essential services of child care and much care of the sick and elderly. Similarly poor material incentives are offered to those who provide opportunities for certain kinds of self-development (e.g., artistic, spiritual) or who do much of the maintenance of private and public spaces. We should therefore not be surprised to find that our system elicits less of this kind of work than is needed, even while it motivates the production of much more than is needed or wanted of some other kinds of output.

Accordingly, another huge industry—but not big enough to soak up all the extra labor—has been created to persuade everyone who has money to keep using it to buy more and more and more—not only the things that are going to make their lives better, but gidgets to create problems, and whizmos to solve the problems created by the gidgets, and more and more and more of this and that and the other. Is this getting us what we want?

What Is It That We Want?

I once asked an anthropologist friend, What are the things that all human beings want? He said security, comfort, honor, and amusement.[3] We have been very successful in achieving three out of these four things.

Regarding *security:* Most of us do not have to worry about where our next meal is coming from, and, while we all do die in the end, it is quite common to live more than seventy years without having to look death in the face.

Among the extraordinary *comforts* we have achieved, productivity growth throughout the eighteenth, nineteenth, and twentieth centuries has made it possibly for the large majority of people in our society to have a quality of furniture and of plumbing—to take just a couple of examples—that would have been the envy of kings and nobles a few centuries ago. That is partly a result of technology, but it also represents the fact that, in terms of access to material things and energy, most of us are in many ways much richer than those ancient aristocrats.

We have done as well in *amusement* as in comfort: Consider our twenty-four-hour-a-day access to books, music, films, and other entertainments.

What about *honor?* I asked my anthropologist friend what he meant by that. He said, "The things that allow us to feel self-respect, and to feel that we are respect-worthy in others' eyes. The things that create a sense of honor differ widely from one culture to another, but the desire for it seems to be a human universal."

In our society, one of the main ways we seek honor is through *success.* For most people, self-respect and the respect of others are closely tied to whether they have achieved success in terms that are recognized by their group. Those terms may differ; they are not the same for academics as for businesspeople; they tend to include some different elements for males, who put more emphasis on sports, and for females, who find it hard to escape from a definition of success that emphasizes physical attractiveness. But all of these definitions of success—which are the underpinnings for respect and honor in our society—are strongly related to what happened to the growth dividend that emerged in the industrial revolution.

We are persuaded, by the producers, that our honor demands that we do a little better than the Joneses (who are working their hardest to stay a little ahead of us), because success has been defined as material success: "He who dies with the most toys wins." And we cannot get off the treadmill, because the money to live a secure, comfortable, amusing, and successful life requires a job; and there will not be jobs if people do not keep buying the output of all those productive workers. This is part of the explanation of why our society has defined honor in terms of success—because that definition keeps us wanting to earn more money, and spend more money, and that keeps the system going. This is the heart of the culture of consumerism.

So here we are, in a situation where society as a whole *must* consume more—even if the things being consumed are not going to add to anyone's well-being or address genuine needs. When anyone suggests that people should resist the pressure to enter the American rat race of earning more in order to consume more, the specter arises of business failures—because producers cannot sell if consumers will not buy. Firm survival depends upon the ability to make profits.

The Appeal of Inequality for a Post-Industrial Society

As a general rule, the highest profits are associated with sales to people who have already satisfied their basic needs. Consider the layout of a modern department store. Which are the items that are most often to be found on the street floor, near the entrance, where they will receive maximum exposure to customers? Perfumes are first; then cosmetics; then jewelry; then "notions" (a catchall term for frippery that is clearly not *needed*). These luxury items get top billing because they have the highest profit margins; there is little pressure to reduce their sale price to anything approaching their cost of manufacture. Profits from sales of nonnecessary things are often dependent on such negative values as waste, conspicuous consumption, obsolescence, fashion, and the creation of false needs.

The relative profitability of the perfumes on the first floor versus the everyday clothes in less accessible parts of the department store illustrates an important reason that *needs* attract so much less attention than *wants*. This is the logical conclusion of a situation—one that must be welcomed by most thoughtful people—in which the productive system has achieved the capacity to do much more than just respond to basic needs. But this desirable outcome is distorted by a number of things. One is the basic nature of capitalism, with the mindless tug of profit—an issue to which we will return. Another is the fact that needs are more evenly spread over the population than money. The essential human requirements for clean water, nourishing food, adequate shelter and warmth, basic health maintenance, and basic education differ relatively little from one person to the next. By comparison, the distribution of wealth and income over most modern populations includes huge disparities. In a market economy, it is the ability to spend money—not needs—that directs the flow of goods and services.

Below is a graph (figure 1) depicting categories of expenditure intended to represent total consumption in an imagined (but not wholly unrealistic) society. The line marked aggregate consumption is intended to symbolize what is actually supplied in this hypothetical society. The vertical axes of figure 1 represent annual expenditure per person. The

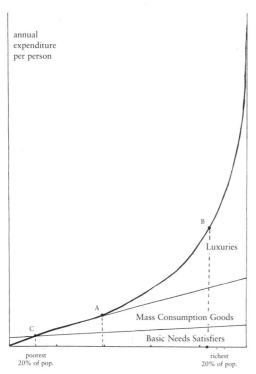

Figure 1. Total household expenditures in a hypothetical economy.

horizontal axis, representing the population, is marked off in quintiles, with the poorest 20 percent of the population represented between the first two marks on the left, and the richest 20 percent between the first two marks on the right.

The aggregate consumption curve indicates the annual cost of purchases by each family. The area below this curve (which could be calculated to tally the society's GDP, or gross domestic product) has been divided into three very general types of expenditure, which will be left quite loosely defined.[4] A line drawn from any point on the aggregate consumption curve straight down to the horizontal axis will indicate the relative proportions of money spent by an individual consumer on the three broad categories of goods. Thus, in this society, an individual at point B spends roughly half her income on luxuries, dividing the rest between basic needs satisfiers and mass consumption goods. An individual at point A buys no luxuries but spends somewhat more on mass consumption goods than on basic needs satisfiers.

The band of basic needs satisfiers drawn at the bottom of the graph is relatively narrow in relation to all the other things people might possibly want. The line separating basic needs satisfiers from mass consumption goods is shown to slope gradually up toward the right, indicating that people with more money are likely to spend more on a better quality of basic needs satisfiers. However, an economic observation that has held over many diverse circumstances suggests that the *proportion* of income spent on basic needs will always decline as income rises.

Obviously, this picture is highly abstract, and the way it looks could change significantly depending on (1) how we define each of the three categories of goods (see note 1) and (2) to what society we are applying it. Regarding point 2, one thing we can infer from the way this particular diagram was drawn is that it depicts a system so productive that even if there were a way to insure (through a guaranteed minimum income, for example) that the market would respond to all of the basic needs in society, there would still be an enormous amount of productive capacity left over. (Of course, this conclusion could change if we changed our definition of basic needs.)

And yet, in fact, even today not all of our society's basic needs are fulfilled. This is because the demand they generate is not always what economists call "effective demand"—that is, it is not backed up by purchasing power (money). Thus, figure 1 shows point C, at which people have fulfilled all of their basic needs, but nothing more. The percentage of the population that exists to the left of point C is living (or dying—quickly or slowly) below the level of basic needs.

The profit motive that is the engine of a capitalist society requires that most productive capacity be utilized and that its output be sold. Figure 1 shows a society in which these requirements, in conjunction with a very skewed distribution of money, result in much more attention being paid to other motives than to the relatively quickly sated basic needs.

This outcome is reinforced, in a much subtler way, by the association of *income* with a *job*. In accepting a job, the individual does not define for him- or herself what needs to be done but instead adopts someone else's definition. The worthwhileness of the work is not the concern of the workers, who are removed from an identification with, or a responsibility for, their output. Nor is the human value of the output a concern of the owners, who often reap the ultimate profits in the impersonal form of stock dividends.

The Downsides of Economic Growth as We Know It

Meanwhile, the forces of economic growth, which have given us so much opportunity for comfort and amusement, are beginning to take back some of our advances in security. One major source of anxiety about the future is the environmental consequences of economic growth. There is much reason to think that we simply cannot continue forever to expand output and consumption: If we do not pull back first, the ecological backlash will stop us in our tracks, through one or more of a variety of unpleasant correctives that nature has up its sleeve. That list of possible correctives includes things like famine and pestilence. Another possibility could be a reversal of the growing labor productivity we have come to take for granted. This productivity has depended in many ways upon the gifts of nature—especially clean air and water and abundant energy and raw materials. If we run into limitations in these areas that we cannot overcome, productivity will decline.

Another aspect of insecurity relates to jobs. A job has tremendous psychic meaning. (I have already noted how we associate ideas of personal success with jobs.) A job is also, for most people, what stands between them and poverty. And the security of that barrier is notoriously diminishing. Unemployment is an alarming, growing issue in much of the world. The increasing skill requirements of modern jobs may be one reason for this; another frequently cited systemic factor is technology, which continues to raise labor productivity, so that the same number of people produce ever more output. If that output cannot be absorbed by some part of the global society, then the number of workers—or at least the number of work hours—has to shrink.

There is a terrible irony in the way this is now being played out. Awareness of the suffering caused by job insecurity and unemployment causes many of the most humane and caring citizens and policy makers to stifle their concern over environmental dangers and join in the call for ever accelerating economic growth. Economic growth may have some problems, but at least—the reasoning goes—it will produce jobs.

It is too often overlooked that *economic growth,* including growth in labor productivity, *does not produce just jobs: it also produces output.* Economic imbalance threatens when demand lags behind output. As global economies are now developing—especially with the growing inequalities that force more and more producers to target the elite "spike" (as in figure 1) of luxury expenditure—it seems unlikely that demand will catch up. Unless some deep changes are made in the way we decide *what* to produce—and for whom—economic growth will continue to produce too much output; and *the result of this kind of growth is more, not less, insecurity and unemployment.*

A turn away from the inegalitarian solution, back toward mass consumption as the sponge for output, would require a different division of wages, away from the morally untenable and economically irrational gap now opened up, especially in the United States, between high-paid, top-level salaries and average wages. This change would not require economic growth—just a return to social norms that were taken for granted only a few decades ago. Unfortunately, although norms do change, and although commercial actors have shown considerable understanding of how to make them change in the direction of greed and materialism, those who would steer social norms in another direction have fewer resources and less agreement on what is really needed.

Thus, the humanitarian would-be reformers, lacking a clear alternative agenda, resist the warnings of the environmentalists and call for more growth. The environmentalists fear that we have already had too much economic growth: too much is being produced, including too many polluting by-products. Technology could reduce the ratio of undesired pollution to desired products, but environmental pessimists doubt that this ratio can continue to be reduced fast enough and long enough to accommodate the growth that the humanitarians say is needed to rid the world of poverty. A few on the humanitarian side chime in with the observation that getting rid of poverty is not enough; we also have to address the causes of envy, which create a perception of poverty in the minds of many in the first world who, by third world standards, are living very well indeed.

Where shall we enter this tangle? I suggest that we return to a look at the growth dividend—what I defined earlier as the increase in output that can be expected in one generation of workers.

Some Ways to Use the Growth Dividend

The economic decline or collapse predicted by the environmental pessimists would be a devastating experience, potentially far worse than the Great Depression of the 1930s. It would also wipe out a great and splendid opportunity, to use a continued growth dividend to make life much better in other ways than by continually increasing consumption. This is, quite possibly, the choice we face:

- either to cling to our current conceptions of growth, increasing, without significantly altering, our present patterns of consumption, until the environmental and/or social backlash forces drastic changes upon us (fortress cities of the affluent are already a symptom of the social backlash);
- or to work out a different kind of growth, in which the dividend of increased potential output from increasing labor productivity is deployed in ways that strengthen the whole fabric of society.

The rest of this section lays out a series of alternative ways to do the latter.

1. An ideal—and idealistic—use of the growth dividend would be to *even out the rewards for work* (what is called "wage compression") so that people who do not have access to the most productive jobs are not so penalized. Thus, the growth dividend would be used (as it was for many years in Sweden) to reduce poverty and/or extreme disparities in wealth. The continuing crisis in the Swedish system points out the difficulties of achieving this in the context of global markets. In that context, the Swedish system went to a possibly unworkable extreme, attempting to work through the market to compress pay differentials further than the market could bear. (At the other end of the scale, the extreme pay and salary differentials now found in the United States also reflect irrational forces, supported by the location of power rather than the logic of the market.)

2. Given the reality that salaries and wages are in important ways qualitatively different from other prices, there may be as-yet-unexplored types of regulations (beyond a minimum wage for the low-pay end or income taxes for the upper end) that could help achieve significant *reordering of the relationships between work and*

income. Any such approach would certainly have to be supported by popular norms that are different from the norms that now have the dominant voice in the United States. We should also be on the lookout for more radical, or more creative, solutions that might look quite unlike what we are accustomed to. A promising example is Time Dollars—a system that allows time and talents, as well as products, that are in surplus in the formal economy to be offered through a different, parallel exchange system.[5]

3. Another, and perhaps more obvious, example of how to use the growth dividend would be to *shorten the work week.* If the average worker can produce 2 percent more output next year than this, why not choose to keep the output steady while adding a week of vacation time to the year? It would, unfortunately, take an enormous amount of social engineering to make that solution work, because productivity growth is not spread evenly in all industries. Nevertheless, this is an option that is being examined closely, especially in several Western European countries.

4. An important possibility is that the growth dividend could be employed to *change the composition of what is produced.* This is an option supported (in different but apparently converging ways) both by environmentalists and by those who are concerned about the moral quality of modern life. Examples of desirable changes in the composition of output could include:

   • *dematerialization*—that is, a continual decrease in the amount of material used per dollar of output. Services are a part of output, and the production and distribution of services are often (though not always) less materials-using than the production and distribution of tangible goods. There is a well-known existing trend in which services account for an increasing proportion of GDP in many countries. Miniaturization, and other aspects of the information revolution, are moving in the same direction.

   • *better design.* Alan Durning, the author of the powerful book *How Much Is Enough?* considers how to reduce some kinds of "personal work," like commuting, and make more time for others, like child care. His proposed solutions focus on urban design, emphasizing ways in which automobile use can be diminished by encouraging public transport and the use of bicycles (including, for the latter, covered roads against rain).[6]

   • *improved working conditions.* The sociologist Fred Block expands this notion to the *quality of economic output* by using that term sweepingly to include satisfying work, economic security, a safe and clean environment, leisure, and the expansion of democracy (in the workplace and in the rest of society).[7]

- *internalizing social and environmental externalities into the sale price of marketed goods*[8]; the resulting changes in prices would spontaneously cause a desirable change in the composition of consumption.

- *a move toward more collective consumption;* that is, relatively more expenditure on public goods—a category that includes things such as public education, public spaces, clean drinking water, and the absence of a cholera epidemic. If some arm of society does not take steps to preserve such public goods, and to prevent them from being appropriated or spoiled by self-interested individuals or groups, you cannot look to the profit-making sector to do this. The traditional way of shifting the composition of consumption toward collective or public goods is to allow more of society's choices to be made and implemented by its democratically elected representatives. This normally implies a larger role for government, including the power to make more of society's spending decisions. The alternative described by John Kenneth Galbraith, of "public squalor, private affluence," was amplified by the Reagan–Thatcher antigovernment mood, whose effects are still with us. It remains to be seen whether ways of strengthening collective consumption that depend more, for example, on philanthropic initiatives than on government will turn out to be acceptably equitable and effective.

5. Most generally, it seems fairly obvious that a reasonable objective is *to shift public and private expenditures, and jobs, out of the areas where they are not socially useful, or produce net harm, and into the areas that combine environmental, social, and personal benefits.* That is obvious and commonsensical, but it goes against much of the structure of our economy. Moving toward this essential change may, for example, require the widespread application of social and "green" business audits, with the results made public. That would have an impact if there also existed a movement of stakeholder groups to monitor the social bottom line of firms, just as stockholder representatives (or U.S. courts, which are strongly biased toward short-term stockholder profit interests) now hold firms accountable for maximizing immediate returns.[9]

(An important note: Any of the just mentioned approaches to changing the composition of consumption could be compatible with an absolute reduction in per capita or total consumption; but none of these logically requires such a reduction.)

Reorienting Business to Human Well-Being

The good news is that there are a lot of smart people trying to interpret laws and institutional changes that will adjust market signals so that the market will be more profitable in producing products that respond to human values, and less profitable in destroying them. An example is the growing trend toward giving workers more ownership of the firms in which they work. That is part of a larger effort to identify all the stakeholders—including neighbors and customers as well as workers—whose lives are affected by what a business does, and to find ways of making those stakeholder voices heard.

There is enough slack in modern production systems so that we can usefully spend the next few decades making them more workable. The immediate challenge is to find ways of offering incentives to businesses to reduce waste, energy use, and overall throughput and to align their goals with more of what will benefit society: increased and more diverse educational opportunities, less violence, more rational land-use patterns, more job sharing, more leisure, and better uses of leisure. Keep your eye on efforts to make companies pay for the damage they do and give a profitability edge to social responsibility; if you want capitalism, as a system, to succeed, such efforts are our best hope.

I have mentioned the idea of making it profitable for firms to engage in socially beneficial production and unprofitable to produce products that tear at the social fabric. Another approach (probably both are needed in concert) is to find ways to get businesses to redefine success, realigning their goals away from the most short-term profits. This is not quite as abstract and daunting as it may seem; in fact, quite a lot of it would be achieved if large corporations, in particular, could be reoriented toward a longer concept of their self-interest. As the time frame lengthens, the self-interest of individuals, corporations, and nations tends to become more similar and to align more closely with environmental health as well. Even such an establishment figure as Michael Porter at the Harvard Business School has made quite radical-seeming suggestions for how to get businesses to identify with social welfare by taking a longer view.[10]

While we are working out the ways to bring the long-term goals that are in everyone's interest into the currently short-term vision of most businesses, we also need to think about the next phase: one in which we accept something like a no-growth, or even a negative-growth, goal for what is currently measured in GDP accounts—even while we would hope to maintain progress in the social accounts. The transition to that longer-term transition is one in which we will have to deeply reexamine consumerism.

A Focus on the Consumer

The economic theory of markets makes a large claim: Whatever else happens to the worker, or to the environment, at least we know that the welfare of consumers is maximized by freely competitive, profit-driven markets.

If what was being produced was what society really wanted—if income were distributed evenly enough so that each family could benefit from the whole society's growth in productivity—and if there were not environmentally destructive consequences to the industrialized world's kinds and levels of production and consumption—the last two centuries' growth in labor productivity would have brought about a state of nirvana. Even if we disregarded the environmental issues, many people would agree that no such blissful state has come to pass. Why not? Here are a few answers.

- If, in the category "consumers," we include people of the future, then we have to be aware that a continued goal of maximizing consumption may be extremely dangerous to future well-being by degrading the resource base on which productivity—hence, the opportunity to consume goods and services—depends. Already, today, we are living in what our parents thought of as "the long run"—and just beginning to discover the environmental price tag on the long-term effects of their economic behavior.
- While, for people whose basic needs are not being met, increased consumption is a very high priority, even that priority is not well served by a system that aggressively markets what the producers want to sell. The most attention is paid to the consumers who have a lot of money to spend on nonessentials; when large corporations address the very poor, they too often focus on the items on which there is a large profit margin (e.g., baby formula) instead of on the staples that serve basic needs. (Remember the layout of the department store.)
- There is strong evidence that, after a reasonable level of basic needs are met, additional goods and services do less and less for consumers' actual well-being. There are ways of spending society's resources that

would evidently contribute much more to general, present, and future well-being than the lifestyles that result from the producers' competitive scramble for profits, achieved by relentlessly pushing a high-consumption mentality.

- If the category "consumers" includes whole persons—not just individuals in one facet of their lives—then we have to face the fact that virtually all consumers are also workers. However, the economic goal of maximizing consumption ignores the actual well-being of workers—even as it also ignores the actual well-being of consumers (focusing, instead, on the single issue of maximizing their market purchases).

A Focus on Values

Consumerism—the linchpin at the heart of the internally consistent, interlocked system that has evolved over the last couple of centuries—is, above all, a set of values. Wherever I, as an economist, have tried to trace causes back to their causes, when I finally come to the font from which the rest all flows, it is always values. Economic systems, as human constructs, are ultimately shaped by what people hold dear—what really matters to them.

The roots of consumerism, I have suggested, are to be found in the past and present needs of the production side of the economy. But the consumerist culture has grown and flourished on its own—with, to be sure, continuous watering from advertising and the media. This culture is hostile to the values that oppose it—thrift and conservation—depicting these as mean and dispiriting virtues.

Values can often be understood in pairs of opposites. Neither member of the pair ever really goes away, but when one is dominant, the other lurks in the subconscious of the national or individual psyche. The particular value pair that is in question here might be given the names *frugality* and *self-indulgence*. The value of frugality was once much acclaimed in the writings of Ben Franklin, or as a part of what it used to mean to be a Yankee, or in the type of thinking that produced the Horatio Alger stories. It has been suffering a downturn for quite a long time, but it is not completely gone. The practical rewards of simplifying life were recently discovered by a New York *grande dame* who announced her discovery that "when you sell the silver, you can get rid of the safe." In my own experience, if you bike to work, you do not need to use the stationary bicycle in the electrically lit and heated health club. While we are in the transition to a transition, it will be critically necessary to find ways to redirect attention to these more traditional values, combating the current ethos in which wasteful consumption is seen as chic and sexy.[11]

When the applications of such practical values constitute proposals for consuming less, such proposals raise ripples of alarm about job loss. The close association of jobs with the health of the macroeconomy, on the one hand, and with the well-being of individuals and families, on the other, is often seen as a justification for a high level of consumption. This logic makes people understandably nervous about the possible economic impact of anticonsumerist values. The key to escaping this close association is to figure out better ways of using our productivity— our growth dividend.

Some of the proposals cited earlier for redeployment of the growth dividend—for example, Durning's ideas about urban design—are dismissed as utopian, with the response that "that would cost too much— we cannot afford it." In effect, response says that we must resist efforts to support an industry designed to improve the quality of life on the grounds that it would increase economic activity. Ironically, the groups espousing that position are often the same ones that protest against decreasing public support for an antisocial industry because to do so would reduce economic activity. We must not hurt the economy by penalizing polluters, or cutting back on the military or support to cigarette manufacture or building roads, according to that view, but we cannot afford to create more jobs in health care, or education, or urban design, or alternative energy sources.

Conclusion

A shift in widely held values is not any easier to achieve than a reorientation of business goals. In fact, the two are very obviously interrelated. And they are intermeshed with all of the other socioeconomic elements I listed to start with, including the especially knotty problem of income depending on jobs, jobs depending on business profitability, and profitability depending on sales to the middle- and upper-income consumers. There is a lot of unraveling to be done here, and little theoretical attention has yet been paid to the macroeconomic consequences of moving toward a world that is organized on the basis of human values more than on values defined as prices.

There are some self-correcting forces in a market economy—the tendency, for example, for prices to rise to reflect growing scarcities. I do not think we can rely on these automatic forces to achieve a profound self-reformation of the system. Any deeper change will require conscious intention and the combined efforts of a great many people bringing to bear the knowledge and understanding of many different ways of thinking and knowing—including a continued cross-fertilization between spiritual and scientific wisdom.

In looking for improvements to our economic system, I do not think we need to start from an assumption of asceticism, denying ourselves things that will really make life better. It may be that, if other changes are not made soon enough, circumstances will eventually restrict our options much more than at present. However, it is just conceivable that humanity will be able to avoid such an outcome by escaping the present high-consumption society and that we can do this simply by discovering and doing what will really give us satisfaction—rather than being pushed by the producing sector into buying things that we do not really want, or that only will give a fleeting pleasure, or that will quickly wear out.

The essential value shift is to redefine our understanding of success and honor. Other, more concrete elements to look for in alternative scenarios include the requirement for a much more even distribution of income and the problem of the tendency for malignant bureaucratic growth.

Our consumer-oriented economic system has created levels of comfort, access to experience and knowledge, release from many kinds of discrimination and bondage, and opportunities for privacy, individual development, and choice that few people would trade for the preindustrial life. The choice, however, is not whether to go back or to go forward; the numbers of people now living on earth simply preclude any possibility of returning to preindustrial conditions. For the world we have inherited, capitalism has strengths that have not been found in any other economic system. However, there are already many different kinds of capitalism, and there may be other variants yet to be developed.

We do not know, yet, whether we can succeed in the only slightly easier task of making our present capitalist system change as needed; nor are we yet at the point where we have to make a transition to a radically new kind of socioeconomic setup. We still have some time, I think, in which to make the transition to that transition.

Notes

1. As I will use this term several times, I should attempt to clarify its definition—not an easy task, in fact, since the term is often employed fairly loosely, and I will not devote the space here that would be necessary to tighten it up significantly.

First of all, when speaking of "mass consumption goods," the term *goods* should be understood to include both goods and services.

Next, to contrast luxury goods and mass consumption goods: Perhaps the outstanding characteristic of the latter is that *they can be, and are, mass produced.* Luxury goods, by contrast, are produced in smaller quantities so as at least to give the appearance of having been customized for a small, elite group of purchasers.

Note that mass consumption goods may include many of the goods and services that are necessary for human survival and may thus be said to serve people's "basic needs." However, many people fulfill their basic needs (for food, shelter, drinkable water, breathable air, etc.) by other means than through a market. It is only when these items enter the market, are mass produced, and are widely available to a class in society that would not be regarded as an elite that they are categorized as mass consumption goods.

For example, in some eastern European cities where air pollution is causing severe respiratory problems, there are breathing clinics where people can stop in for some breaths of purified air. What had once been a freely available satisfier of basic needs is, thus, turned into a luxury good. (Some of the clinics require steep fees; others are available only to an "elite," defined in terms of a statement of medical need.) If these clinics were to become highly accessible throughout the population, at a widely affordable price, we could then categorize their products as mass consumption goods—or even (if they were actually necessary to a reasonable life span) as basic needs satisfiers.

2. See, for example, Shirley P. Burggraf, *The Feminine Economy and Economic Man: Reviving the Role of Family in the Postindustrial Age* (Reading, Mass.: Addison-Wesley, 1997). I would note that while I applaud Burggraf's brilliant analysis of the problem, I hope very much that we can find a better solution than the one she proposes. Perhaps the most striking thing about her chilling proposal is that, when you think of it, you realize that it is no worse than our present system.

3. Allan Hoben, professor emeritus of anthropology at Boston University. Personal communication, 1993.

4. See note 1. This figure has greatly simplified the reality of purchasing patterns. Among the things it does not show are the fact that some people will cut back on basic needs to purchase not just mass consumption goods but also some luxuries; and, more generally, the fact that the division between the three categories of expenditures is not so smoothly predictable on the basis of income. The purpose of the figure is to illustrate, in a very general way, some of the realities of spending patterns in a rich Western society. Its applicability to the United States or to other particular societies can be argued. Does all of the population have all of its basic needs met? If so, the aggregate consumption curve should not go beneath the line for basic needs. How much of the total spending of society is done by the small elite

to the right of point C? Perhaps the aggregate consumption curve should continue higher (to indicate more spending by this group), or stop shorter, etc.

5. For more information on this system, a good source is the Maine Time Dollar Network web page, which provides a general introduction to Time Dollars and links to related web pages. The address is http://www.gwi.net/mtdn.

   Other sources include a book by Edgar Cahn and Jonathan Rowe: *Time Dollars: The New Currency That Enables Americans to Turn Their Hidden Resource—Time—into Personal Security and Community Renewal* (Emmaus, Pa.: Rodale Press, 1992). Shorter pieces may be found in *Yes! A Journal of Positive Futures*, Spring 1997 (P.O. Box 10818, Bainbridge Island, WA 89110, 206-842-0216), especially "Money with Care Built In" by Jonathan Rowe (pp. 20–23) and "Money That Builds Connection" by Edgar Cahn (pp. 24–25).

6. Alan Thein Durning, *How Much Is Enough? The Consumer Society and the Future of the Earth* (New York: W.W. Norton, 1992). See also Durning, *The Car and the City* (Seattle: Northwest Environment Watch, April 1996).

7. Fred Block, *Post-Industrial Possibilities: A Critique of Economic Discourse* (Berkeley: University of California Press, 1990).

8. Externalities are costs that are borne by someone different from the economic actors who are setting the prices. Examples of what happens when externalities are internalized follow:

   - If the producer were obliged to pay the full cost of the losses in health or amenities resulting from pollution associated with production, the forces of supply that do half the work of price setting would push the price of a good higher.
   - Similarly, if consumers had to pay the full price of disposal or other consumption-related costs that are now largely externalized onto society, they would buy less of a product.

   Either of these shifts would result in decreased sales of an item after significant externalities are internalized to either the buyer or the seller. This would shift the composition of what is exchanged in the market away from those goods.

   If people bought fewer items whose externalities had been internalized, what would they buy more of instead? The answer is: either items whose externalities had not yet been internalized or items that do not have negative external effects. The latter is, obviously, the preferred result.

   Another real possibility, however, is that if all social and environmental costs of production and consumption were internalized, people would buy less of everything, because so many things have externalities (and would with the externalities internalized cost more) that it would be impossible to exclude all of these items from one's shopping list; thus, in effect, the cost of living would go up. The antidote to this possibility is technological progress that would reduce undesirable side effects of consumption and production. Economists generally hope that the rise in prices, and the consequent change or reduction in total purchases, caused by internalizing externalities would sharply spur this kind of technological progress.

9. Significant progress is being made in several of the areas mentioned here, especially by the Coalition for Environmentally Responsible Economies (CERES). CERES's work toward the creation of a Global Reporting Framework (GRF), organized in concert with interested NGOs throughout the world, as well as with accounting firms and other stakeholders, has resulted in a blueprint for auditing the social and environmental impacts of business. During the trial phase (1999–2000), a number of Fortune 500 corporations, as well as other smaller firms, have committed to the data collection and reporting required for the GRF. Monitoring groups are mobilizing to ensure that the results reach the public and promote constructive feedback to the participating firms. (For more information, contact CERES at 11 Arlington St., 6th floor, Boston, MA 02116-3411; www.ceres.org.)

10. Michael Porter, "Capital Choices: National Systems of Investment," in Neva R. Goodwin, ed., *As If the Future Mattered: Translating Social and Economic Theory into Human Behavior* (Ann Arbor, Mich.: University of Michigan Press, 1996).

11. The New American Dream is an organization that has been created specifically to hold up to Americans the possibility of discovering a good and fulfilling life through the values of frugality and responsibility, as opposed to self-indulgence and me-first-ism. It may be found on the Web at www.newdream.org.

Christian Responses to Coercion
in Population Regulation

Susan Power Bratton*

But the midwives feared God, they did not do as the King of Egypt commanded them. . . .

—Exodus 1:17

As world population passes the six billion mark, meeting human resource needs worldwide becomes ever more complex and daunting. Both the wealthier industrial nations and the two-thirds world struggle to provide enough food for all their new families, adequate schools for their children, and dependable employment opportunities for their young adults. Ecologists, economists, government officials, educators, and health professionals concerned about public welfare make projections about the future of agriculture in Kenya, the possibility of disease outbreaks in megacities, and the impact of immigration on unemployment in the United States. Meanwhile, environmentalists still discuss the population explosion, the notion of a sustainable planet, and the future of "spaceship" earth. Beneath all these global models and earthwide concerns, however, are millions of families and individuals who make both reproductive and economic decisions. All too often the rights and concerns of parents and children as individuals get lost in the

*Susan Power Bratton is Lindman Chair of Science, Technology, and Society at Whitworth College in Spokane, Washington.

rhetoric of planetary ecological or economic crisis. Conversely, the voices for "the family" or "the economy" sometimes disguise the negative impacts that speedy demographic change, particularly population increases, can have on individuals.

Considering the universal nature of changes in human demography and of ever increasing competition for access to natural resources, the Christian theological community has had relatively little to say about the ethical and cosmological issues that a rapidly expanding human population presents. Many seminaries offer courses on sexuality or family life, while ignoring the environmental and economic implications of human demography. Few of the theological volumes available on sexual ethics analyze regional cultural and social class differences that may influence decisions concerning child bearing and child rearing. Often the perceptions and feelings of real women lie buried in abstract cases about ideal families. Conversely, the extensive feminist literature often neglects the potential benefits of discussing women's rights issues in religious or theological terms. Christian theology is still a fruitful ground for community-based dialogue seeking both constructive responses to day-to-day social conflicts and loving approaches to the support and nurturing of human life. Biblical ethics and even contemporary theological approaches, such as liberation and feminist theology, remain vastly underutilized in the construction of Christian responses to demographic change.

The purpose of this chapter is to investigate one major ethical issue arising from attempts to regulate population growth: the potential use of coercive methods to change birth rates or other population characteristics. Almost any program that attempts to influence reproductive rates through reward or punishment should at least be suspect and may raise questions about the rights of parents, the rights of children, and possible injustice to one gender or the other or to a specific social class or ethnic group. When is a government- or community-sponsored family planning or population regulation program coercive, and when does it unduly influence the decisions of families and specifically of women? How far should a government go in influencing birth rates or family size? When does government interference violate human rights?[1] I will use biblical texts to reflect on the ever present tensions between seeking the common good and honoring individual autonomy. I will also employ biblical standards for social justice and concern for the poor and marginalized as a benchmark to guide humane social responses to some of the major questions confronted by today's potential parents, especially women.

Coercion and Reproduction

Although we think of "population regulation" as a twentieth-century passion, through history human societies have implemented various forms of social sanctions and positive and negative incentives in attempts to suppress or stimulate population growth. Bioethicists most often discuss the legislation and policy of official governments, but human cultures have long established social conventions, such as the ostracism of unwed mothers or the encouragement of late marriage, that tend to limit population size. One of the common mistakes made by Christians is to assume that pro-natalist programs, those encouraging population growth via increased birth rates, are de facto pro-child and pro-woman. Such programs may actually result in serious degradation of the health and well-being of women and children and can undermine family integrity. The communist-era restrictions on contraceptive availability in Romania, for example, left many abandoned infants in orphanages. The Third Reich's rewards for German women who had large families, such as the gold "mother's cross," valued only Aryan offspring. Children of other ethnic backgrounds starved behind the barbed wire of concentration camps and suffered horrid, untimely deaths in gas chambers. The Nazi state encouraged the select SS "Aryan supermen" to father children, even out of wedlock, thereby placing eugenics well above family values.[2] Nazi pro-natalism was anything but pro-child or pro-family.

Coercion's ancient roots and its negative consequences for women are reflected in the Hebrew scriptures. The story of baby Moses floating down the mighty Nile in a little papyrus basket, made watertight by a layer of pitch, is a narrative about coercive population regulation. The Hebrew people sought sustenance in Egypt during famine and stayed to become slave labor. When a new pharaoh ascended the throne, he worried about their growing numbers and legislated a reduction in their head count by ordering the Hebrew midwives to safely deliver female infants and kill all male newborns. Exodus 1:17 reports: "But the midwives feared God; they did not do as the king of Egypt commanded them, but they let the boys live."[3] The midwives Shiphrah and Puah risked not just their community standing and employment but also their lives in resisting the king. Their "fear of God" does not merely suggest that they were terrified of being struck down by divine power, it implies a deep understanding of the role of God in the community and as Creator and Sustainer.

When summoned by the Pharaoh to explain their disobedience, the midwives said that they had had no opportunity to dispatch the boys "because the Hebrew women are not like the Egyptian women; for they are vigorous and give birth before the midwife comes to them." Exodus 1 and 2 are highly affirming of the oppressed Hebrew women. The names of the midwives come from *spr* meaning "beautiful" and *p'h* meaning "splendor" or coming from a word associated with childbirth.[4] The statements of the midwives declare the strength of the Hebrew women—which is the fortitude not just to bring a child into the world unassisted, but also to resist Pharaoh himself. Today's midwives still struggle against unjust government policies, but the battle is to protect infants, to bring reproductive health care to women, and to provide safe and optimally timed childbearing.

Moses's mother and sister execute the second major act of women's resistance when they hide the newborn boy for three months and then in desperation set him afloat on the Nile. Exodus 2:4 notes: "His sister stood at a distance to see what would happen to him." When the baby is rescued at the command of Pharaoh's daughter, it is Moses's sister who offers to fetch a wet nurse, who is, of course, the baby's birth mother. The Egyptian aristocrat recognizes that the baby is probably Hebrew but demonstrates no concern whatever for her father's dictates. The narrative becomes a conspiracy of women, who join together to defend the child. His race, ethnicity, and possible humble origin do not matter.

Advocates of family planning or of women's health services often avoid the story of baby Moses, because the narrative deals with the evils of an anti-natalist policy. The Book of Exodus as a whole, however, is about oppression, liberation, and the eternal presence of God in universal terms, and its lessons extend far beyond ancient ethnic discrimination.[5] Exodus begins in a socially disastrous circumstance in which an unjust ruler quickly adopts coercive population control as a means of reducing the political power of a subject ethnic group. As Donald Gowan and other commentators have pointed out, Exodus 1–2 are also the first chapters of the Bible in which God is initially absent and in which he openly intervenes on behalf of his people.[6] In the concluding chapters of Genesis (with the exception of one verse) and in the first chapters of Exodus, "Up to the story of the Lord's revelation to Moses the name *'Elohim* and not the Tetragrammaton is used."[7] Further, the slaves have lost both the name of God and their history, just as in coercive population regulation they are losing their future. The suffering of the Hebrew people originates in an external evil, as "they are not accused of having done anything wrong."[8] Pharaoh's population policy oppresses them for who they are, not for any realized crimes or insurrection. During the Exodus, they escape Egypt, learn God's name, and recover their eternal future.

The biblical narrative has several elements that are common to many cases of state-sponsored coercion, including both forced abortion and sterilization and complete lack of accessible reproductive health services. First, the major motive for Pharaoh's policies is ultimately economic. Pharaoh wishes to both control the Hebrews and exploit their material value. He could expel them from the kingdom, but he chooses instead to drain their potential strength into the Egyptian economy while reducing their social power by limiting the number of potential "alien" warriors under his rule. The new king irrationally fears the non-Egyptian peoples in the land yet values them for building new cities. He attempts to manage their demography to the advantage of the kingdom—as his dynasty. It should be no surprise that when Moses the man comes to ask for the release of his people from bondage, which would remove the Hebrews as a political threat, yet another pharaoh says no, and the plagues begin. Although Pharaoh's actions are anti-natalist, his motives are little different from those of an economist or government official who suggests that large human populations are desirable because they provide cheap labor and stimulate economic growth by increasing markets, yet no provisions are made for the increased infrastructure, including educational and health care facilities, that rising populations require.

Second, Pharaoh's program is anti-ethnic and differentiates between "native Egyptians" and "resident aliens." Both pro- and anti-natalist policies may discriminate on the basis of ethnicity, origin, or social class as well as economics. If gynecological and health services are more available to women from the wealthier and upper classes, both maternal survival and child survival will be better for the wealthy than for the poor. Although not as direct a form of oppression as Pharaoh's legislated infanticide, economic discrimination in health care allocation can still throw infants into the Nile.

The ancient Egyptians also practiced gender selection in their program to reduce the numbers of surviving infants. Coercive contemporary population programs usually eliminate female fetuses by abortion or encourage the abandonment of female infants. The "one child per family" policy in China, for example, has led to both higher birth rates for males and a high proportion of female and disabled children in orphanages. The Egyptian preference for girls may be rooted in high value in the workforce combined with their lower potential to organize a martial threat. Young women could also serve as wives or, more likely, as concubines and household servants. Entering Egypt because of a famine, the patriarch Abraham said to his wife Sarai: "I know well that you are a woman beautiful in appearance; and when the Egyptians see you, they will say 'This is a wife'; then they will kill me, but they will let you live" (Genesis 12:12). Abraham lies and says Sarai is his sister, therefore, and "when the officials of Pharaoh saw her, they praised her to Pharaoh. And the woman was taken to Pharaoh's house" (Genesis 12:15). Pharaoh's protection of female infants originated in his view of them as chattel. The Egyptian program that set young Moses afloat was capable of destroying the Hebrew family and presumably the system of inheriting family names and property, while leaving all Hebrew beauty (spr) in Egyptian hands.

In Exodus, the effective resistance to coercion arises from the women's sphere. Midwives are still the major figures at the "birthing stool," and in defense of both mothers and children resist the power of the state. In a time when the name of the true God is forgotten, God "helps Israel in Egypt invisibly yet effectively by means of the mid-wives."[9] Exodus 1:20 reports that "God dealt well with the mid-wives"—an affirmation of both their social role and their faithful execution of their duties. Today, medicine and government often partially remove birthing from the female spheres of influence or control, and it is common for medical professionals or the managers of a population or family planning–oriented program to be from one social class or ethnic group, while the people the program is supposed to serve are from another. As Exodus clarifies, coercion may be reduced or thwarted when the "reproductive professionals" truly reflect the interests of the women and the families the professionals serve.

The behavior of Pharaoh's daughter could be regarded as a flippant disregard for the wishes of her father, but it might be better interpreted as the appropriate response of a woman of integrity. The text stresses that in looking at the abandoned and crying child, "She took pity on him" (Exodus 2:6). She also appears to have recognized the hand of divine providence when she accidentally encounters the basket in the reeds, and in finding the child, she realizes he has become "hers." In paying wages for a wet nurse, she takes responsibility for a child she clearly knew was born outside her social class. Biblical scholar Umberto Cassuto notes: "The keyword in this paragraph [about the rescue of Moses] is *child*, which occurs seven times. Also emphasized in the paragraph, by being used three times are the words *took, called, went* and *suckled*." Both Pharaoh's daughter (who remains forever unnamed) and the woman of Levi "take" the child.

The entire Book of Exodus is, in fact, a discourse on providence and on divine love. For the Egyptians, the Nile was both a deity and the source of the nutrients and water necessary to agriculture. The midwife brought (male) children into the world, and the river brought Pharaoh to his throne and a mighty civilization into the land. Ironically, when Pharaoh orders the newborns killed on the birth stool or thrown into the Nile, he forces the role of executioner onto those who give or protect life. The Egyptian aristocrat who found the baby while bathing in the Nile names the boy Moses, which means "son" or perhaps "begetting a child" in Egyptian. The word is often found in the Egyptian names of deities, such as Thutmose.[10] The author of Exodus, however, interprets the name in Hebrew (and it is unlikely the Egyptian princess spoke that language). Exodus 2:10 reports: "She named him Moses, 'because,' she said, 'I drew him out of the water.'" When the river gives the baby to a princess, it is as if the life-giving powers of nature are working in concert with the will of the Hebrew God. The Hebrew simultaneously suggests the drawing out of the water (as in birth), and the child's destiny as a deliverer, who would draw his people "from a sea of servitude."[11] Through the remainder of Exodus, as a divinely ordained event, the river and the land of Egypt give up the children of Israel and return them to their rightful parent(s) 'Elohim. Ironically, today we need to not only consider the protection of mothers and children but also the mighty river itself. No longer those of a deity, the Nile's waters deposit its nutrient-rich silt behind human-made dams, and its life-giving powers continue to shrink as more and more people crowd both its mouth and its headwaters.

One of the great evils pursued by Pharaoh was his attempt to turn the figures and forces that protect and nurture life into death-dealing sources of oppression. State-sponsored population programs that remove the wonder of birthing from women, or deny the importance of the "child," or force women to abandon their children show no respect for the Creation and lack a deep understanding or "fear" of God's relationship to humanity. The names of the midwives provide a clue to avoiding coercion—that birth should always be spr, or beautiful. Today, too many offspring, especially in regions where there is still high infant and child mortality, may be as much of a tragedy as too few.

In summary, the story of baby Moses identifies the following as dangerous and ungodly sources of coercion:

- irrational fear of an underclass, resident aliens, or ethnic minorities;
- economic policies that only drain labor from the people and offer no fair reward or adequate sustenance for work contributed to national welfare;
- loss of respect for the life-giving and life-preserving powers of the

Creation, and for God-given providence—turning the Creation from life supporting to death dealing;

- removal of women's or family control or influence in birthing;
- valuing one gender more than the other, in terms of labor, military participation, or inheritance of material goods or political leadership;
- denial of the beauty and sacred nature of human reproduction; and
- denial of the importance of allowing all people to honor their past, fully live in the present, and seek their future with God.

Pro-Natalism and the New Testament

In the time of Caesar Augustus, the Roman empire pursued pro-natalist policies, at least for Roman citizens. The emperor was worried that falling birth rates, particularly in urban Rome, would understaff his armies. The solution was incentives to procreate, such as tax breaks for married men, and legislation to curtail social practices that limited birth rates, such as herb-induced abortions.[12] At the same time, even citizen children could be sold into slavery, child prostitution was common, and parents frequently "exposed" female and handicapped infants, leaving them to die unattended.[13] Augustus still had to struggle to feed sprawling, slum-filled Rome and favored a mix of policies that kept Roman-led legions stationed in grain-rich regions like Palestine, while worrying little if proletariat children were at risk of starving to death. Not just Jewish society, but Mediterranean cultures in general, continued to value women primarily for their role as wives who produced male heirs and as concubines and servants. In ancient Palestine, the barren woman or the childless widow might easily find herself out on the street with no honest source of income and no family support.

In contrast to Augustus, Jesus of Nazareth was a quiet king, whose empire required no particular lineage of soldiers to defend it, and whose banquet table could seat guests from any social class or nation. The teachings of Jesus counter the programs of Pharaoh and Caesar alike because they propose that children (and women) are more than economic entities, characterized primarily by their material value to the husband and head of the family or to the state. Christ writes a new chapter in holy history by his affirmations. When the people are bringing their children to Jesus and the disciples rebuke these persistent children's welfare enthusiasts and their ill-mannered offspring, Jesus says: "Let the little children come to me, and do not stop them; for it is to such as these that the kingdom of heaven belongs" (Matthew 19:14). He also reaches out to those when royal Rome would have abandoned, including widows, deathly ill little girls, and women who were not from the house of Israel, such as the Samaritan divorcee at the well and the Syrophonecian woman whose daughter had a demon (probably seizures). Both Jesus and the disciples affirm those who may not have been reproductively fit, including the woman with the flow of blood who was ritually unclean (Matthew 9, Mark 5, and Luke 8) and the Ethiopian eunuch who became the first evangelist to Africa (Acts 8). When the early church called the first deacons, their job was ministering to widows—many of whom had lost their "reproductive value" in a male-oriented society. Even for the least of these, and the most unprocreative of these, it is worth before God and not material value that counted in the early "Jesus movement."

The New Testament thus provides further instruction in avoiding coercive reproductive policies:

- the ill and infirm deserve loving care, not social exclusion;
- children must be valued as beloved of God and not as sources of economic return or even as members of a particular ethnic group;
- reproductive fitness or the ability to procreate and have a family does not convey status in the Kingdom, nor is parenting necessary to provide honorable service to God;
- reproductive policies or social norms should not exclude individuals from access to necessary daily resources and sources of financial support; and
- forced reproduction or heavy emphasis on the bearing of sons is not consistent with the goals of Christ's Kingdom.

It should be reiterated that these roots of coercion potentially apply to both pro-natalist and anti-natalist policies.

Are Some Population Regulation Programs Coercion Free?

It may seem like a great leap from the ancient world to our contemporary issues, but social fears and economic concerns still dictate national policies. We should first ask, however, whether some types of state- or nonprofit-sponsored population regulation or family planning efforts are strictly noncoercive and inherently just. The first question concerning justice is, therefore: Is medical advice concerning contraceptive options or the means of delaying or spacing childbirth or limiting family size likely to cause injury to anyone, or to limit anyone's life choices and personal autonomy? Although almost any type of medical or social service may have unexpected negative outcomes, such as a reaction to prescription drugs, for the vast majority of its practitioners, family planning protects life and the integrity of individuals. Proper birth spacing is, in fact, extremely important to infant survival and maternal health. Further, limiting family size often provides more resources for each child. In rural regions and urban slums, younger children from large families are at higher risk of disease and malnutrition. Preventing pregnancy when a woman is in ill health or has a new baby may save the life of the mother or the child. We have to conclude that providing family planning assistance to socially responsible adults is far more likely to be beneficial than detrimental, both to the individuals seeking the services and to their communities.

Even if a family planning program is medically safe and largely socially positive, however, there may be a point at which it unduly influences the behavior or philosophy of its participants. Almost any effective medical or social service that offers advice or contraceptives, or that simply cares for pregnant women, will probably influence its clients' or patients' views. The potential role of advertising is an exceptionally sticky issue at the beginning of the twenty-first century, when mass communications, in the process of selling jeans, cars, and beer, also sells sexuality as the primary source of self-fulfillment. If one watches television frequently, one might become convinced that material goods and an enticing personality are all there is to life. We need, therefore, to ask: Is using advertising or education in a population program that is intended to encourage smaller families unethical?

In turning to the Scriptures, a first conclusion is: Advertising and education are not wrong, since that's how the church was established. Spreading "good news" is blessed by God. The question is more one of content and intent. Advertising to encourage sexual and reproductive responsibility may, in fact, help to counter the "bad news" that surrounds us. Advertising is coercive if it magnifies imagined threats or encourages irrational fears. Conversely, it is responsible if it warns of real dangers, in an honest and thought-provoking way. Advertising is also coercive if it belittles women or children, denigrates an ethnic group, or derides the joys of parenting. Perhaps the worst advertising denies the Eternal and encourages us to acknowledge only the present moment by finding our fulfillment solely in the material. Appropriate advertising concerning family planning looks beyond the question of the safe sexual act and, while acknowledging the rights and privacy of individuals, links individual choices to responsibility to others, including the responsibility of parenting wisely. If advertising and education help to reduce sexually transmitted diseases, improve infant survival, encourage responsible parenting, or help poorer families to space their births and keep their children in school, then they aid the economically and socially disadvantaged, particularly children at risk. As long as sexual irresponsibility and disregard for women and children are not simultaneously encouraged, advertising concerning family planning meets New Testament mandates for service to others.

Can Positive Incentives Be Wrong?

In the United States, everyone is a potential game show contestant and expects to receive "free offers" of prizes in the mail. We may find it hard to believe use of positive incentives to encourage socially beneficial actions could be wrong, when we so often use them to encourage selfish actions and irrelevant behaviors.[14] It is worth noting that along with the Kingdom, Jesus offered his disciples troubles and persecution, homelessness and tribulation, and they still went for the package. Incentives can enslave the will and blind one's spiritual vision, or they may result in injustice to someone else. Having to make one's own choices and bear the consequences is usually the healthiest long-term route to mature—and righteous—behavior.

The list of positive incentives that might encourage reproductive responsibility is long, but at the first level an incentive is merely something for nothing or for a minimal cost, such as:

- free sex education;
- a free or subsidized family planning service, providing hormone implants or other contraceptives;
- free or subsidized health services, with added attractions such as free check-ups or free infant care; and
- free sterilization.

The next level of incentives is positive social acknowledgment without substantial material rewards. Although society is usually inclined to commend people for doing more instead of doing less, it is still possible to commend couples with small families or to give certificates to high schoolers who vow celibacy until they are ready to enter a mature relationship with someone else and to responsibly fulfill their duties as a potential parent. It is also possible to give certificates or other social affirmation to individuals completing health education programs.

The third level of positive incentives is to provide extra resources for participating in family planning programs. These may include:

- cash incentives or other material rewards for sterilization;
- cash incentives or tax breaks for small families;
- better housing or jobs for small families;

- scholarships or other educational support for children from small families;
- special professional or economic opportunities for people who delay marriage or childbearing; and
- group incentives, such as offering a village a new tractor, pump, or secondary school if it reduces overall birth rates.

The New Testament expresses a deep concern for meeting the day-to-day needs of everyone, including the poor and socially marginalized. Providing for universal education and health care, however, is not the same as providing a reward for a specific type of behavior. Distributing positive incentives means making a resource differentially available. We should therefore distinguish between the first level of positive incentives, which are merely free or easily available services and thus serve as "entitlements," and the third level of incentives, which are contingent on behavior and social cooperation and whose distribution will not be equal to everyone who is interested. Further, giving one family a large apartment means giving another, perhaps larger family, a small one.

The New Testament is often skeptical of our usual cultural practices concerning material rewards. Jesus preaches to those who have the least in this life and declares in Mark 10: 43–44 to his disciples: "But whoever wishes to become great among you must be your servant, and whoever wishes to be first among you must be the slave of all." Incentive systems may in fact be coercive if they are more attractive to the poor, or encourage people to sacrifice their "family care values" for a material possession, such as a transistor radio. This, in fact, has been one of the major criticisms of "sterilization for cash." Because the family is large enough, or because only one son will be able to inherit the farm, or because one wishes to share resources with one's neighbors is a better motive for seeking sterilization.

A biblical comment on the distribution of rewards may be found in the story of Hannah, who through long years of barrenness was still well treated by her husband. In 1 Samuel, when her spouse Elkanah is attending the sacrifice at Shiloh, he allows his two wives to join in the feast, and "he would give portions to his wife Peninnah and to all her sons and daughters, but to Hannah he gave a double portion, because he loved her, though the Lord had closed her womb" (1 Samuel 1:5–6). Elkanah continues to dedicate himself fully to her in her childless state and gives her the better piece of meat at the feast. Hannah eventually conceives and bares the prophet Samuel, whom she dedicates to God. Although Elkanah's cultural expectation was "many children," the Bible mentions him as someone who loved and cared for his wife despite her "inadequacies." Differential rewards, for bearing either too many children or too few, should never lead to social or family rejection.

In dealing with positive incentives, we can identify a number of ethical concerns. First, an incentive program that denies children from large families or lower birth order access to resources will harm individuals who did not choose their own social circumstances. Thus, special education benefits for small families are undesirable if they leave some children out. If possible, reward systems should directly influence adult behavior concerning contraception and should not target specific social classes or ethnic groups. In countries where access to contraceptives has been limited by poor health services, the first step is not to establish an incentive system but to make contraceptives available to all couples who wish to use them.

After adequate health services are available, a reward system may be justified if there are competing social values that discourage use of family planning. A program of rewards should not undermine either the religious beliefs or the social identity of cultural groups that resist reducing family size. In regions, however, where a large number of sons has traditionally conveyed increased social status or has been considered economically advantageous, a reward system may help to compensate for loss of status or income through reduction in family size.

One cannot consider "benefits" to be true incentives if they are forced (e.g., if the only way to receive a food allotment is to be sterilized). Incentives offered under false pretenses are also questionable. If a nation is offering free sterilization to those who already have children, it would be dishonest to avoid informing the participants that surgical procedures will make them *permanently* sterile. Conversely, participants in incentive programs may report that they are taking birth control pills when they are actually selling them or throwing them away. A man may also wait until he has fathered a large family and then go and collect his incentive, so he collects a reward while contributing nothing to shared community welfare. For this reason, incentive programs are often a more effective means of reducing family size when they encourage permanent or long-term methods and provide greater rewards for couples who stop at a smaller family size.

Although a perfectly equitable incentive system is probably unattainable, some approaches are better than others. India, for example, has provided cash payments. Such efforts often pay more to men (in the 1970s India offered $20 for a vasectomy and $2 for a tubectomy) and are therefore gender discriminatory.[15] If surgical procedures are not carefully executed, the more difficult tubectomy presents more risk of infection or of injury that disables her for work. Since approved methods of contraception for women continue to exceed those available for men, the family may benefit from the reward, while the woman bears the risk. Incentive systems should consider gender equity, in terms of both potential dangers and benefits.

A major difficulty with financial incentives is that they are obviously more attractive to the poor. One means of reducing injustice, however, is to reduce child mortality so that if a poor man or poor woman is sterilized after having a small number of children, those offspring are more likely to survive to adulthood. If incentives help to equalize surviving family size between social classes or ethnic groups, they are probably relatively justly distributed. If incentives—like Pharaoh's program, which intended to draw all the reproductive resources to the side of the dominant cultural group—provide larger families for the wealthy or dominant social classes, or provide better opportunities for their children, they should be considered very ethically suspect.

Negative Incentives

If entitlements or positive incentives do not work, or if a community or government thinks it cannot afford them, negative incentives may be employed—that is, incentives under which the person who does not cooperate suffers some harm or loss. A government with may also decide that participation in a family planning, program, like conscription in the military, is the responsibility of all good citizens and will be equitably required. When the government has little inclination toward democracy and protection of human rights and fears population growth, the resulting population programs can be draconian.

Common types of negative incentives or social coercion include, at the least coercive level, all forms of casual suggestion or societal unwillingness to aid the large family. Social ostracism is commonly practiced when families are too large (or in some cultural settings, too small). Even such subtle conventions as the two- or three-bedroom apartment norm in public housing may discourage large families (or alternatively family peace).

At the next level are standard charges for more children, including, commonly, additional school fees, and in some cases extra taxes. The economies of urban industrial societies, where there is limited housing, and legislation to discourage or prohibit employing minors, often strongly discourage large families. Rural economies where children may be gainfully employed in farm work often have the opposite impact.

At the most coercive level are strict regulation of family size or of the

number of live births per woman. Governments may fine or jail offend-
ers or remove privileges or economic benefits, such as vacations, job
promotions, or salary increases. Governments may also force sterilization or contraception on families who do not want it, or in the most
severe cases force abortion on women who already have one or two
children. Through the centuries, many societies have killed or abandoned infants and even young children when families became too large
or a new infant was born too soon after an older sibling. Although
Pharaoh ordered the male infants slain, even today a preference for sons
adds to what is known as "the girl child problem," in which female
children are more likely to be abandoned or left without resources in
times of shortage.[16]

Negative incentives present all the difficulties positive incentives present—and more. First, the impact of such social controls may fall on the
innocent. Even if a father is fined for having too many progeny, the
anger and frustration of the guilty parents are likely to be transferred to
his children. If larger families pay more tax, the youngest children, who
are likely to be shortchanged in terms of resource access anyway, may
suffer loss of educational support or even starvation. Negative incentives may gravely worsen "the girl child problem." If a family is expecting to be fined for having two children and one is female, she may find
herself on the orphanage doorstep, or worse, be abandoned at a bus
station or quietly smothered.

All coercive methods of encouraging reduced birth rates restrict the
rights and choices of the individual and therefore conflict with basic
values in Euro-American culture, where privacy and the sexual choices
of the individual are coveted symbols of personal freedom. Champions
of civil liberties who have attempted to eliminate government interference in sexual acts between consenting adults also condemn such overt
invasions of privacy as forced sterilization. Even in the civil liberties–
conscious United States, however, government agencies and public institutions have sterilized the mentally retarded and residents of psychiatric
wards without their consent. Physicians have self-righteously performed unnecessary hysterectomies on women from racial or ethnic
minorities, in some cases without properly obtaining the patient's permission, or without clearly informing her that she will be sterile after-
ward.[17] Again, the ability to procreate is a blessing from God and
should not be taken from anyone without her or his consent.

Coercive methods of population regulation potentially represent a
collision between two competing sets of values: the integrity of the individual and the safety of the community. Following contemporary ideals
that consider the rights of the individual inalienable, we could consider

violating the right of privacy only when, as in the case of war, the community is really under a life-or-death threat. This is a lesser-of-two-evils argument, which requires that several criteria be met before coercive means of regulating population can be considered.

First, coercive methods should be employed only where famine is a serious threat or mortality due to *widespread* malnutrition is occurring, and the crisis is directly related to population levels and not to some other factor such as lack of land for peasant farming. In other words, some restriction of individual reproductive options is appropriate where mortality or great suffering will result if populations continue to expand. The threat must be proven—the projections of environmental alarmists are not by themselves adequate to justify coercion. Second, before forced participation in birth rate reduction programs is considered, food, health care, educational opportunities, and other resources must be relatively equitably distributed within the society. Coercion is not appropriate where malnutrition is strongly related to social class, where a majority of the populace is not well fed, or where socioeconomic problems independent of population growth are causing shortages. In the latter case, just distribution of resources would be the first step in relieving famine. Further, large families are often a response to poverty and economic insecurity. Forced sterilization and required use of contraceptives are not solutions to social insecurity originating in unjust land tenure systems or the growth of urban slums because of unemployment in rural regions. Third, an attempt must have been made to make family planning services widely available and people have been unwilling to use them or to reduce family size. Often, provision of education for women, operating in concert with improved medical services, will be extremely effective in reducing birth rates *without any form of coercion*.[18] Fourth and last, other methods of encouraging reduced birth rates must have already been seriously attempted and have failed. These methods might include honest persuasion, such as advertising campaigns that tout the benefits of adequate birth spacing.

Few societies worldwide presently meet these four criteria. Many nations with large rural populations, for example, have not adequately coped with land tenure issues. Throughout much of the third world, rural and "underclass" populations still do not have easy access to family planning. Millions of women worldwide would like to increase the interval between births, delay the birth of their first child, or reduce total family size and cannot get the medical assistance they desire. Further, although almost all traditional societies successfully manipulate individual choices like age of marriage through social sanctions and economic control, both democratic and authoritarian governments have found that even the less politically active or less educated portion of the populace will reject direct government interference in their reproductive lives. Customs concerning marriage and childbearing often have long cultural histories and the support of most of the adults and leaders within a specific cultural group. A majority of people will adhere to those social norms because they are what they have always been taught is right. They expect, and receive, local community affirmation for their compliance with local community values. When a government initiates coercive methods, those methods may be no worse than the social sanctions or punishments imposed by the community for violation of traditional mores. Government coercion, however, may represent quick and unwanted change, may be viewed as external interference without community consent, and may create deep values conflicts. If, for example, the government tells a Muslim that large families are bad and unpatriotic, while his upbringing and religious culture have taught him that large families are good, the Muslim will probably respond by placing the values of a centuries-old religion and of his cultural heritage above those of a changeable and politically suspect contemporary bureaucracy.

When Paul Ehrlich, in *The Population Bomb,* suggests that the use of coercive methods is ethically justified by present *world* population growth rates,[19] he is very much the biological and social expert taking a parental attitude toward those who do not understand the dangers of accelerated demographic change. As K.S. Shrader-Frechette has commented, he is unwilling to wait for democracy to operate.[20] Coercive methods pose the dangers of racism, sexism, classism, and culturism. Social sanctions have always been most effective when a majority of the populace agree to their use and willingly participate both in obeying the rules and in enforcing them. Although China, where central government control is exceptionally strong, has been successful in using negative incentives in controlling fertility, India has found population politics can influence elections. In 1975, Prime Minister Indira Gandhi used powers of emergency rule to upgrade a fertility reduction campaign. Some Indian states used strong negative incentives such as withholding food rations from those with more than three children or threatening salary loss to those who did not participate in a sterilization program. Actual cases of coercion, in concert with rumors of further planned coercive actions, helped to defeat Mrs. Gandhi's government in an election in 1977.[21]

In order to preserve democratic processes and protect human rights, we can add a further criterion that should be met before coercive methods and negative incentives are used: The implementation of coercive methods must result and should not depart from democratic processes. The populace should favor and support any social sanctions or negative incentives used (just as one would hope the citizens of a nation would all vote in favor of legislation aimed at controlling drunk driving). Legislation should be instituted by vote, and public programs should be approved by elected bodies or other appropriate governmental units, such as village elders.

Christians should also be aware that socially oblivious state-sponsored programs that employ coercion in family planning undermine the value of religious teaching in society and the religious rights of individuals and families. Such programs are often based in very material or coarsely utilitarian philosophies about the meaning of Creation and deny the significance of the birth of each individual human. It is extremely difficult for military dictatorships and government bureaucracies to comprehend the Scriptural meaning of "made in the image of God," much less to demonstrate any "fear" of God whatsoever. Although Christians may not agree with all the theological arguments other world religions pose to support large (or small) family size or their marriage customs, we should appreciate their ethical wisdom, their deep concerns for the welfare of humanity, and the wellsprings of their cultural heritage. We should also remember how much we value our own freedom to voice our religious thoughts and views.

With advances in birth control technology producing more long-term contraceptive strategies that are reversible, the options for temporary restrictions on reproduction are increased. Not all possible strategies are equally desirable, however. In actually establishing a program of mildly coercive methods or negative incentives (assuming famine or similar crisis), the following are possible ethical criteria for selecting techniques.

First, methods of birth control or fertility regulation causing possible death or injury of either parents or children should be avoided. From a Christian perspective, practicing infanticide to avoid mortality of older children or adults is trading one life for another. Prevention of conception should be emphasized as the appropriate means for reducing total population size. Historically, many cultures have abandoned certain types of individuals during periods of resource limitation or social stress. Those left without support have included female infants, infants in general, younger children, older women, and the aged in general. The modern trend has been to consider all individuals of equal worth—an idea that Christianity has forwarded. The New Testament continues to encourage us to think of the Kingdom as available to all, and not as favoring male over female, married over unmarried, virility over age, or maturity over youth.

Second, use of reversible contraceptives (particularly long-term contraceptives) is to be favored over other methods of reducing family size. This preserves the greatest number of options for the individual couple, should family or economic circumstances change. (Voluntary sterilization might be encouraged as an alternative for couples who *know* they have as many children as they want.) We need to recognize that reproductive and other life or biological processes, and the divine providence on which they are based, are not totally subject to our less than divine control, and that even a well-thought-out "family plan" may need to be changed.

Third, if legislation establishes family size criteria for participation in certain programs (including the receipt of rewards), these should ultimately be based on number of children surviving the high-risk early childhood years rather than on the number of births. Family limitations based solely on number of infants delivered are usually prejudiced against the poor, who are more likely to lose a young child. Allowing a woman who has lost a child to bear another will help couples to achieve the goal of having progeny available to assist them in their old age. When Jesus raised the son of the widow of Nain from the dead and gave the man back to his mother, the miracle served the economic needs of an individual who had no other source of support.

Fourth, application of family size criteria should be equal for all social strata in a society. If large families are discouraged, the rich, well-educated, and fully employed should be under the same sanctions as the poor, uneducated, and underemployed. The Scriptures respect the desire for children, and so should we. No one who wants a family should be excluded because she is poor.

Fifth, programs should avoid strategies that result in discrimination against female children, especially de facto infanticide of girls. In societies that have traditionally favored sons, trying to achieve very small family sizes or extensively utilizing abortion to limit family size is likely to result in gender bias. Jesus ministered to women and considered them fully worthy of hearing his teaching and inheriting the Kingdom.

Sixth, negative incentives should not directly or indirectly harm children—for example, subjecting them to parental rejection or denying them places in school. Nor should they harm women or the economically disadvantaged. Elkanah's love should be our model in distributing the best portion at the feast.

Seventh, forced sterilization and abortion are invasions of the individual person (or, put in Western terms, violations of individual rights and of privacy) and are not acceptable. (There is a potential counterargument here: Sterilization after the production of several offspring has not prevented the individual from reproducing and does little personal damage [there is less risk for men than for women], whereas in a time of severe food shortages, for example, bearing more children will result in further resource stress and further mortalities. Protecting privacy and reproduction potential does not take precedence over preventing further deaths. For many Christians who believe the fetus is a person, however, this argument cannot apply to forced abortion.)

Eighth, legislating the age of marriage or requiring a certain economic status prior to marriage is acceptable for delaying the onset of childbearing, providing restrictions do not force women to bear children outside of legally sanctioned matrimony and do not discriminate against the poor or against particular ethnic groups. Delaying marriage is acceptable if the delay is equally shared among all individuals of reproductive age and if it does not force unmarried pregnant women into a marginal social status.

Ninth, since most contraceptive techniques are not completely dependable and fertility varies from person to person, establishing patterns of childbearing based on absolutes is highly undesirable. A woman who has no intention of conceiving another child may become pregnant despite her best efforts. It might be appropriate to require a woman who has had three children to utilize a hormonal implant in order to receive an extra food allotment. It would not be appropriate, however, to take the food allotment away permanently if contraception failed and she delivered a fourth child, because both the health of both the mother and her children would be likely to suffer as a result. Restrictions based on absolutes generate social cruelty and force women into very specific childbearing roles, which they may or may not be willing or able to fulfill. Again, we humans have many options for the right management of our reproduction and of the Creation, but, as the author of Ecclesiastes has so aptly noted, wickedness of others and chance disasters happen to all of us, despite our good efforts.

Tenth and last, eliciting individually responsible behavior and giving couples an opportunity to regulate their own fertility is always superior to government-enforced efforts to maintain a specific reproductive standard. The Scriptures present a repeated call for individuals not just to lead righteous lives themselves but to properly care for their families. The Bible repeatedly encourages us to share what we have with others, and this should be applied both to the resources immediately needed to raise children and to our relationship with Creation as a whole.

In summary, we can conclude that negative incentives and coercion in general are ethically problematic and should be employed only as a last resort. Even in crisis, however, some methods preserve the rights and options of the individual better than others, and programs must be carefully planned to avoid doing more harm than good. Very few nations, if any, presently meet the criteria for utilizing strongly coercive tactics. The controversial Chinese program, which has been critiqued internationally for its human rights violations, is, on one hand, just because it (1) originated in response to famine and to genuine resource shortages, (2) was preceded by an attempt to make family planning assistance widely available, (3) is based in a society that is attempting to share resources relatively equitably, and (4) pursues family size restrictions that are free of social class distinctions. As successful as it has been in terms of stabilizing population growth, however, the program violates several of the ethical criteria listed above. It has (1) been established without consideration of democratic processes, (2) greatly restricted individual reproductive options, (3) violated individual privacy, (4) forwarded fatal gender discrimination, and (5) forwarded abandonment and mistreatment of some children. If one considers abortion to be killing a person or violating the rights of a child, then the Chinese program trades one life for another. Further, even though the government does not sanction it, the program has encouraged the killing of newborns. The Chinese program (through the 1980s and 1990s) has repeatedly failed to protect the rights of both pregnant women and unwanted infant girls.[21] The Chinese have done very well in making the program socially equitable and poorly in preserving the integrity of the individual. We have to ask ourselves if this is the necessary result of letting population growth get out of hand, and if widespread mortality from famine can be resolved only with the sacrifice of individual rights. The weaknesses of the Chinese program make a timely attempt at a distributive justice approach even more attractive and suggest that the nations of the developing world should take population issues very seriously as they struggle to obtain better lives and greater freedoms for their peoples.

Conclusion

Well-planned and thoughtfully executed public health and family planning programs that forward the welfare of individual women and children fall well within the biblical mandates to care for our neighbors. Coercive programs should serve only as a last resort, and in a world where many women still do not have the means to obtain regular medical care, much less contraceptives, the best and most ethical strategy remains increased provision of education, family planning services, and access to medical professionals. Christians have neglected to address many of the ethical issues raised by demographic change yet will find that the Scriptures and available ethical and theological models may be readily applied to dialogues concerning the most critical issues.

Notes

1. See my book *Six Billion and More: Human Population Regulation and Christian Ethics* (Louisville, Ky.: Westminster/John Knox Press, 1992), for a more complete discussion of the history of population issues. Some other Christian authors who have written on population include James Nash, *Loving Nature: Ecological Integrity and Christian Responsibility* (Nashville: Abingdon Press, 1991); Sean McDonagh, *To Care for the Earth: A Call for a New Theology* (Santa Fe: Bear & Co., 1987); Loren Wilkinson, ed., *Earthkeeping in the 90s: Stewardship of Creation* (Grand Rapids, Mich.: William B. Eerdmans, 1991). See also James Martin-Schramm, "Population Policies and Christian Ethics," in Laurie Ann Mazur, ed., *Beyond the Numbers: A Reader on Population, Consumption, and the Environment* (Washington, D.C.: Island Press, 1994).
2. Claudia Koontz, *Mothers in the Fatherland: Women, the Family, and Nazi Politics* (New York: St. Martins Press, 1987).
3. All biblical quotes in this chapter are from *The New Oxford Annotated Bible*, eds. Bruce Metzger and Roland Murphy (New York: Oxford University Press, 1994).
4. U. Cassuto, *A Commentary on the Book of Exodus* (Jerusalem: Magnes Press, Hebrew University, 1983), pp. 13–14. See also Martin Noth, *Exodus: A Commentary* (Philadelphia: Westminster, 1962), p. 22.
5. African-American exegesis has an extremely rich tradition of interpreting Exodus for its liberation context. See, for example, volumes by such authors as James Cone, Peter Paris, Gayraud Wilmore, and Dwight Hopkins and the great outpouring of recent work on womanist themes.
6. Donald Gowan, *Theology in Exodus: Biblical Theology in the Form of a Commentary* (Louisville: Westminster/John Knox, 1994), pp. 1–7.
7. Cassuto, *A Commentary*, p. 16.
8. Gowan, *Theology in Exodus*, p. 6.
9. Noth, *Exodus*, p. 24.
10. John Durham, *Word Biblical Commentary: Exodus*, vol. 3 (Waco, Tex.: Word Books, 1987), p. 17; and *The New Oxford Annotated Bible*, p. 71.
11. Cassuto, *A Commentary*, p. 21.
12. John T. Noonan, *Contraception: A History of Its Treatment by Catholic Theologians and Canonists* (Cambridge, Mass.: Harvard University Press, 1986).
13. Michael J. Gorman, *Abortion & the Early Church: Christian, Jewish &*

Pagan Attitudes in the Greco-Roman World (Downers Grove, Ill.: Intervarsity Press, 1982).

14. The sections on positive and negative incentives are a revision and update of materials from Bratton, *Six Billion and More,* pp. 176–85 in chapter 10. Further justification for the arguments used here may be found in that volume, particularly in the preceding chapter.

15. Leslie Corsa and Deborah Oakley, *Population Planning* (Ann Arbor: University of Michigan Press, 1979), pp. 253–54.

16. Among Christian organizations, World Vision has prepared some powerful materials on the "girl child problem." For example, World Vision, *Faces of Poverty and Population: The Washington Forum* (Monrovia, Calif.: World Vision, 1993).

17. Thomas Shapiro, *Population Control Politics: Women, Sterilization, and Reproductive Choice* (Philadelphia, Pa.: Temple University Press, 1985).

18. A recent volume with several readable review chapters on women's desire for family planning is Laurie Ann Mazur, ed., *Beyond the Numbers: A Reader on Population, Consumption, and the Environment* (Washington, D.C.: Island Press, 1994).

19. Ehrlich, *The Population Bomb* (New York: Ballantine Books, 1968).

20. Shrader-Frechette, *Environmental Ethics* (Pacific Grove, Calif.: Boxwood Press, 1981). See especially her chapter on "lifeboat ethics."

21. One of the first scholars to adequately present the Chinese violations of women's rights was Betsy Hartmann in *Reproductive Rights and Wrongs: The Global Politics of Population Control and Contraceptive Choice* (New York: Harper & Row, 1987).

To Protect the *Whole* of Creation

Bruce Babbitt*

I began 1995 with one of the more memorable events of my lifetime. It took place in the heart of Yellowstone National Park, during the first week of January, a time when a layer of deep, pure snow blanketed the first protected landscape in America. But for all its beauty, over the previous sixty years this landscape had been an incomplete ecosystem; by the 1930s, government-paid hunters had systematically eradicated the predator at the top of the food chain: the American grey wolf.

I was there on that day, knee deep in the snow, because I had been given the honor of carrying the first wolves back into that landscape. Through the work of conservation laws, I was there to restore the natural cycle, to make Yellowstone complete.

The first wolf was an alpha female, and after I set her down in the transition area, where she would later mate and bear wild pups, I looked through the grate into the green eyes of this magnificent creature, within this spectacular landscape, and was profoundly moved by the elevating nature of America's conservation laws: laws with the power to make creation whole.

I then returned to Washington, where a new Congress was being sworn into office, and witnessed power of a different kind.

---

*Bruce Babbitt is the secretary of the U.S. Department of the Interior.

Attack on Water, Lands, Creatures

First I witnessed an attack on our national lands, an all-out attempt to abolish our American tradition of public places—whether national parks, forests, historic sites, wildlife refuges, or recreation areas. Look quickly about you, name your favorite place—a beach in New York harbor, the Appomattox Courthouse, the great Western ski areas, the caribou refuge in the Arctic, the pristine waters off the Florida Keys— each of these places is at risk. In October 1995, in the *Denver Post,* the chairman of the House Subcommittee on Public Lands estimated that his committee may have to close more than 100 of the Park Service's 369 units. In these times, it seems that no part of our history or our natural heritage is sufficiently important to protect and preserve for the benefit of all Americans.

Next I witnessed an attack that targets the 1972 Clean Water Act, the most successful of all our environmental laws. Until that act passed, slaughterhouses, pulp mills, and factories from Boise to Boston to Baton Rouge spewed raw waste into our waterfronts. Yet twenty-three years later, as I visited America's cities, I saw that act restoring those rivers, breathing new life into once dead waters. I saw people gather on clean banks to fish, sail, swim, eat, and live. I saw that, as the act helps cities restore our waters, those waters restore our cities themselves. And then I saw Congress rushing to tear that act apart.

But finally, more than any of our environmental laws, the act Congress had most aggressively singled out for elimination—the one that made Yellowstone complete—is the 1973 Endangered Species Act. Never mind that this act is working, having saved 99 percent of all listed species; never mind that it effectively protects hundreds of plants and animals, from grizzly bears to whooping cranes to greenback cutthroat trout; never mind that it is doing so while costing each American sixteen cents per year. For Congress—while allowing for the above charismatic species, plus a dozen other species good for hunting and fishing, plus, just for good measure, the bald eagle—can find absolutely no reason to protect all species in general.

Who cares, members of Congress ask, if the spotted owl goes extinct? We won't miss it, or, for that matter, the Texas blind salamander or the kangaroo rat. And that goes double for the fairy shrimp, the burying beetle, the Delphi sands flower-loving fly, and the virgin spine dace! If they get in our way, if humans drive some creatures to extinction, well, that's just too bad.

The Values of Children

Congress members are not, however, the only Americans who have expressed an opinion on this issue. I read an account of a Los Angeles "Eco-Expo" in April 1995, at which children were invited to write down their answers to the basic question: "Why save endangered species?" One child, Gabriel, answered, "Because God gave us the animals." Travis and Gina wrote, "Because we love them." A fourth answered, "Because we'll be lonely without them." Still another wrote, "Because they're a part of our life. If we didn't have them, it would not be a complete world. The Lord put them on earth to be enjoyed, not destroyed."

Now, in my lifetime I have heard many, many political, agricultural, scientific, medical, and ecological reasons for saving endangered species. I have in fact hired biologists and ecologists for just that purpose. All their reasons have to do with providing humans with potential cures for disease or yielding humans new strains of drought-resistant crops, or offering humans bio-remediation of oil spills, or thousands of other justifications of why species are useful to humans. But none of their reasons moved me like the children's.

These children are speaking and writing in plain words a complex notion that has either been lost or forgotten or has never been learned by some members of Congress, and indeed by many of us.

The children are expressing the moral and spiritual imperative that there may be a higher purpose inherent in creation, demanding our respect and our stewardship quite apart from whether a particular species is or ever will be of material use to humankind. They see in creation what our adult political leaders refuse to acknowledge. They express an answer that can be reduced to one word: values.

A Sacred Blue Mountain

I remember when I was their age, a child growing up in a small town in northern Arizona. I learned my religious values through the Catholic Church, which, in that era, in that Judeo-Christian tradition, kept silent on our moral obligation to nature. By its silence, the church implicitly sanctioned the prevailing view of the earth as something to be used and disposed of however we saw fit, without any higher obligation. In all the years that I attended Sunday mass, hearing hundreds of homilies and sermons, there was never any reference, any link, to our natural heritage or to the spiritual meaning of the land surrounding us.

Yet outside that church I always had a nagging instinct that the vast landscape was somehow sacred, and holy, and connected to me in a sense that my catechism ignored.

At the edge of my hometown, a great blue mountain called the San Francisco Peaks soars up out of the desert to a snowy summit, snagging clouds on its crest, changing color with the seasons. It was always a mystical, evocative presence in our daily lives. To me that mountain, named by Spanish missionaries for Saint Francis, remains a manifestation of the presence of our Creator.

That I was not alone in this view was something I had to discover through a very different religion. For on the opposite side of the blue mountain, in small pueblos on the high mesas that stretch away toward the north, lived the Hopi Indians. And it was a young Hopi friend who taught me that the blue mountain was a truly sacred place.

One Sunday morning in June he led me out to the mesa-top villages, where I watched as the Kachina filed into the plaza, arriving from the snowy heights of the mountain, bringing blessings from another world. Another time he took me to the ceremonials where the priests of the snake clan chanted for rain and then released live rattlesnakes to carry their prayers to the spirits deep within the earth. Later I went with him to a bubbling spring, deep in the Grand Canyon, lined with *pahoes*— the prayer feathers—where his ancestors had emerged from another world to populate this earth.

By the end of that summer I came to believe, deeply and irrevocably, that the land, and that blue mountain, and all the plants and animals in the natural world are together a direct reflection of divinity, that creation is a plan of God, and I saw, in the words of Emerson, "the visible as proceeding from the invisible."

Genesis and the Deluge

That awakening made me acutely aware of a vacancy, a poverty amidst my own rich religious tradition. I felt I had to either embrace a borrowed culture or turn back and have a second look at my own. And while priests then, as now, are not too fond of people rummaging about in the Bible to draw our own meanings, I chose the latter, asking: Is there nothing in our Western Judeo-Christian tradition that speaks to our natural heritage and the sacredness of that blue mountain? Is there nothing that can connect me to the surrounding Creation?

There are those who argue that there isn't.

There are those industrial apologists who, when asked about Judeo-Christian values relating to the environment, reply that the material world, including the environment, is just an incidental fact, of no significance in the relation between us and our Creator. They cite the first verses of Genesis, concluding that God gave Adam and his descendants the absolute, unqualified right to "subdue" the earth and gave human beings "dominion over the fish of the sea, and over the fowl of the air, and over every living thing that moveth upon the earth." God, they assert, put the earth here for the disposal of humans in whatever manner they see fit. Period.

They should read a few verses further.

For there, in the account of the deluge, the Bible conveys a far different message about our relation to God and to the earth. In Genesis, Noah was commanded to take into the ark two by two and seven by seven every living thing in creation, the clean and the unclean.

He did not specify that Noah should limit the ark to two charismatic species, two good for hunting, two that might provide some cure down the road, and, say, two that draw crowds to the city zoo. He specified *the whole of creation.* And when the waters receded, and the dove flew off to dry land, God set all the creatures free, commanding them to multiply upon the earth.

Then, in the words of the covenant with Noah, "when the rainbow appears in the clouds, I will see it and remember the everlasting covenant between me and all living things on earth." Thus, we are instructed that this everlasting covenant was made to protect the whole of creation, not for the exclusive use and disposition of humans, but for the purposes of the Creator.

Now, we all know that the commandment to protect creation in all its diversity does not come to us with detailed operating instructions. It is left to us to translate a moral imperative into a way of life and into public policy—which we did. Compelled by this ancient command, modern America turned to the national legislature, which forged our collective moral imperative into one landmark law: the 1973 Endangered Species Act.

Lost Values, Fragmented Creation

The trouble is that during the first twenty years of the Endangered Species Act, scientists and administrators and other well-intentioned people somehow lost sight of that value—to protect the *whole* of creation—and instead took a fragmented mechanistic approach to preserve individual species. Isolated specialists working in secluded regions waited until the eleventh hour to act, then heroically rescued species—one at a time.

Sometimes the result was dramatic recovery, but often the result was chaos, conflict, and continuing long-term decline. In the Pacific Northwest, for example, the spotted owl was listed even as federal agencies went forward with clear-cutting. Efforts to save the alligator proceeded even as the Everglades shriveled from diverted waters. California salmon runs were listed even as water users continued to deplete the spawning streams.

It is only in the last few years that we have recovered, like a lost lens, our ancient religious values. This lens lets us see not human-drawn distinctions—as if creation could ever be compartmentalized into a million discrete parts, each living in relative isolation from the others—but rather the interwoven wholeness of creation. Not surprisingly, when we can see past these human-made divisions, the work of protecting God's creation grows both easier and clearer.

It unites all state, county, and federal workers under a common moral goal. It erases artificial borders so we can see the full range of a natural habitat, whether wetland, forest, stream, or desert expanse. And it makes us see all the creatures that are collectively rooted to one habitat, and how, by keeping that habitat whole and intact, we ensure the survival of the species.

For example, in the Cascades, the spotted owl's decline was only part of the collapsing habitat of the ancient forests. When seen as a whole, that habitat stretched from Canada to San Francisco. Not one but thousands of species, from waterfowl of the air to the salmon in their streams, depended for their survival on the unique rain forest amid Douglas fir, hemlock, and red cedar. Our response was the president's [President William Jefferson Clinton] Forest Plan, a holistic regional agreement forged with state and local officials and the private sector. Across three state borders, it keeps critical habitat intact, provides buffer zones along salmon streams and coastal areas, and elsewhere provides a sustainable timber harvest for generations to come.

That's also the lesson of Everglades National Park, where great flocks of wading birds are declining because their shallow feeding waters are drying up and dying off. Only by erasing park boundaries could we trace the problem to its source, hundreds of miles upstream, where agriculture and cities were diverting the shallow water for their own needs. Only by looking at the whole South Florida watershed could state and federal agencies unite to put the parts back together, restore the severed estuaries, revive the park, and satisfy the needs of farmers, fishermen, ecologists, and water users from Miami to Orlando.

This holistic approach is working to protect creation in the most fragmented habitats of America, from salmon runs in California's Central Valley to the red-cockaded woodpecker across southeastern hardwood forests, from the sand hill cranes on the headwaters of the Platte River in Central Nebraska to the desert tortoise of the Mojave Reserve. I'd like to say that the possibilities are limited only by our imagination and our commitment to honor the instructions of Genesis.

Let Us Answer

Yet more and more, the possibilities are also limited by some members of Congress. Whenever I confront some of these bills that are routinely introduced, bills sometimes openly written by industrial lobbyists, bills that systematically eviscerate the Endangered Species Act, I take refuge and inspiration from the simple written answers of those children at the Los Angeles expo.

But I sometimes wonder if children are the only ones who express religious values when talking about endangered species. I wonder if anyone else in America is trying to restore an ounce of humility to humankind, reminding our political leaders that the earth is a sacred precinct, designed by and for the purposes of the Creator.

I got my answer in the fall of 1995.

I read letter after letter from five different religious orders, representing tens of millions of churchgoers, all opposing a House bill to weaken the Endangered Species Act. They opposed it not for technical or scientific or agricultural or medicinal reasons, but for spiritual reasons.

And I was moved not only by how such diverse faiths could reach so pure an agreement against this bill, but by the common language and terms with which they opposed it, language that echoed the voices of the children.

One letter, from the Presbyterian Church, said: "Contemporary moral issues are related to our understanding of nature and humanity's place in it." The Reform Hebrew Congregation wrote: "Our tradition teaches us that the earth and all of its creatures are the work and the possessions of the Creator." And the Mennonite Church wrote: "We need to hear and obey the command of our creator who instructed us to be stewards of God's creation."

And suddenly, while reading these letters, I understood exactly why some members of Congress react with such unrestrained fear and loathing toward the Endangered Species Act. I understood why they tried to ban all those letters from the Congressional Record. I understood why they are so deeply disturbed by the prospect of religious values entering the national debate. For if they heard that command of our Creator, if they truly listened to His instructions to be responsible stewards, then their entire framework of human rationalizations for tearing apart the act comes to nought.

I conclude by affirming that those religious values remain at the heart of the Endangered Species Act, that they make themselves manifest through the green eyes of the grey wolf, through the call of the whooping crane, through the splash of the Pacific salmon, through the voices of America's children.

We are living between the flood and the rainbow: between the threats to creation on the one side and God's covenant to protect life on the other. Why should we save endangered species? Let us answer this question with one voice, the voice of the child at that expo who scrawled her answer at the very bottom of the sheet: "Because we can."

Part V
CONCLUSIONS

Earth Literacy for Theology

Mary Evelyn Tucker*

I recently spent a week in the Sonoran desert of southern Arizona near Tucson. This was my first extended exposure to the topography of a desert bioregion. And what a remarkable tapestry it is. It is a landscape that invites silence and contemplation, as the Desert Fathers of the early Christian church recognized. It is also an atmosphere that opens to experimentation and dreaming. These insights may be the roots of a new kind of earth literacy, which is required of both science and religion in our times. Without this the future may be very bleak, indeed. For consumption and population problems will not be solved by formulas or forecasting alone, but through new worldviews created by the emerging multiform field of ecology. This is a field that calls us, as never before, beyond "disciplinalotry" to new modes of interdisciplinary discourse. Hence the significance of a volume such as this, drawing on both science and religion for new paradigms of thinking through our current impasse. In short, as ethicist Daniel Maguire has said, "If current trends continue, we won't."[1] And it is increasingly evident that consumption and population are at the heart of the matter. Unbridled greed of the first world combined with unrestrained human growth worldwide is a formula for disaster.

Let us return to the desert for some pathways through this Scylla and Charybdis. The desert offers us the metaphor of space from which to step back and reconceptualize the problem. It is from this kind of perspective

*Mary Evelyn Tucker is a professor of religion at Bucknell University. Earlier versions of this essay were published in *Teilhard Perspective* (December 1995–June 1996) and *Earthlight* (Summer 1996).

that we can begin to see how to re-envision ourselves as earthlings dependent on nature for life and sustenance. From this space we can begin to self-correct this negative feedback loop of hyperconsumption and population growth. I am not offering answers here—only a space and perspective from which some functional possibilities may arise.

This space is at the heart of earth literacy, which I am defining as scientific understanding of natural systems in conjunction with human perception of social systems. The integration of these two is the goal of earth literacy in formulating new human–earth relations. It is simply a recognition that we humans are a subsystem of the earth systems, that we are not anthropocentric beings but "anthropocosmic" realities.[2] In other words, as Thomas Berry has put it, the earth is primary and we are derivative.[3] The octopus of growing consumption and population trends threatens all life forms on the globe unless we reorient our fundamentally anthropocentric perspectives and worldviews. This will require creative rethinking in the midst of mounting complexity and information overload. It is not more knowledge that we seek but rather, dare I say it, "wisdom"—from both scientists and theologians.

We in the field of religious studies, for example, have been somewhat effective in describing divine–human relations and human–human ethics, but we have been deficient in articulating appropriate human–earth relations. This requires a return to the desert. For Jewish biblical thought and Christian theological thought, the "desert" has traditionally symbolized a perspective that entailed a predominantly human-centered anthropology and a divine reality that is transcendent. We need to ask where is the earth in these reflections? What does the desert say to us?

As our friends told us in Tucson, the desert bioregion seems to attract dreamers, and thus Arizona has its share of unusual peoples and communities. On the one hand, there are retired people seeking peace, quiet, and sun. On the other, are young people looking for space, experimentation, and bonding. In between are several extraordinary houses of reflection and contemplation—inviting people into a space of rich silence punctuated with the rhythm of ritual.

In this mode of silence, one can begin to drink in the desert landscape. It fills one's soul with the extraordinary diversity of life in the midst of such sparseness. There are, for example, over two thousand plant species in the Sonoran desert alone—the richest desert of its kind among our four deserts in the U.S. West. Seeing the huge two-hundred-year-old saguaro cactus standing as sentinels over this desert life is moving beyond words. Touching a pink flowering caterpillar cactus, absorbing the green of a paloverde bark, stepping on the sandy seabed soil, climbing to an ancient cave overlooking an arroyo canyon literally stops one in one's tracks. And then one can be even more attentive to the roadrunner at one's doorstep, the gellner quail at the feeder, the javelina eating cactus outside the window. From a space of contemplation one is more attentive to life, its textured variety of forms, and its interactive ecological systems.

The desert thus calls forth silence and *contemplation* so as to see life again in fresh ways. It also evokes *experimentation* and *dreaming* to new imaginative horizons. Arizona contains numerous examples of this, from astronomical observatories to contemporary constructed communities to futuristic experiments. Star gazing there is a way of looking into our stellar past, our light years of evolutionary history to the formation of galaxies and planets. Moreover, the constructed communities, such as Paolo Soleri's Arcosanti in Arizona, are ways of modeling in the present more sustainable communities that seek a balance of human population and consumption needs. Finally, a unique experiment for the future has emerged in the Arizona desert, namely, Biosphere 2.

Past perspectives are crucial, and present models are indispensable, yet future forecasting is most unpredictable, especially without a broad framework of earth literacy. This experiment of Biosphere 2 has received a great deal of attention in both news media and journals. Controversial even in conception, it has been equally controversial in its results. It may, in fact, be seen up until now as one of the noblest failures of our time. With $200 million of Bass family money, a three-acre enclosed space was built to house various biomes, or complex ecological systems. The hope was to reconstruct natural ecosystems in a controlled environment and see what could be learned for life here or for future life in space. The bioregions built replicated a tropical rain forest, a savannah, an ocean, a marsh, an agricultural area, and a desert.

In viewing this, I was amazed to see the supreme irony of this experiment in reproducing nature. Imagine which of these complex ecosystems is not functioning. In fact, it is dead—the desert of Biosphere 2 amidst the fecundity of the actual Sonoran desert of Biosphere 1, the earth. The complexity of this sparse bioregion was defeated by too much water in the artificially enclosed space. Such marvelous irony for a system that perhaps had some promise for greater ecological understanding but in its overall conception has something of a technofix mentality behind it. When I suggested this "technofix" idea to our tour guide, who exuded unbridled enthusiasm for the project, she said, "It's not a 'technofix,' it's an 'ecofix.'" That may be precisely the point.

While we no doubt have to learn to live in greater harmony with nature's ecosystems, if we imagine ourselves as ecoengineers replicating and controlling these processes blindly or without a sufficiently broad context, we will be operating out of a misplaced paradigm of human arrogance. Is it not presumptuous to imagine that we are "in control" of nature, about which we still know so little? Granted, the experiment of Biosphere 2 was to help us to know more about nature with the assumption we can "fix" the damage we have already done. But, still, something is missing from this picture. As Thomas Berry has said, "Humans cannot create a blade of grass and there is liable not to be a blade of grass unless humans learn to protect and to foster it in a context of humility not manipulation." The framework needs to shift so as to envision ourselves as part of, not apart from, the earth.

A great deal is already being done within such new frameworks, namely, the promising technologies of sustainability that have emerged from highly creative individuals such as Amory Lovins on energy, John and Nancy Todd on agriculture, and Susan Murcott on water treatment. In addition, it was announced just after this conference that Columbia University's Lahmont-Doherty Earth Observatory will take over the operations of Biosphere 2 for a five-year period. This will give Biosphere 2 new directions as a center of environmental study associated with a major research university.

One more example from the desert may be sufficient to suggest another model for earth literacy. That was a conference that brought together twenty people from the natural and social sciences and the humanities to discuss the evolutionary process and its implications for a new paradigm for redirecting and healing human–earth relations. The conference was wonderfully successful for many reasons, perhaps most important in that it allowed sufficient space for exercise and play, conversation and silence, food and sleep. And all of this in a desert context. It was especially significant in allowing the space for a respectful sharing of expertise from various disciplines including the arts, astronomy, physics, chemistry, biology, geology, and the humanities.

This suggests to me that careful listening, from a place of silence, to the voices of the human and nonhuman world is a key to this earth literacy that we are groping to create. The listening begins and ends with allurement and attunement. As Thomas Berry has said, "We have become autistic to the voices of the Earth." A new attunement is required to recover these multiple powers of observation. Initially, attunement may demand atonement; ultimately, it can result in at-one-ment.

As we begin to listen from a space of open reflection, it is likely that we will experience a grieving at what has been lost from our planet already and at what will disappear in our lifetime. As we acknowledge the sorrow or even the despair this entails, we can begin to locate a more realistic position from which to foster life and creative balance. This unacknowledged well of despair may prove to be a constructive source for action in spite of the apparent odds against achieving new modes of ecological equity and equilibrium in human–earth relations.

Hopefully, we are poised in a new phase of a self-correcting loop, as someone in the field of cybernetics might say. Indeed, the world historian William McNeill suggests in his book *The Human Condition* that all civilizations come to points of equilibrium between microparasites (forces of internal decay) and macroparasites (forces of external destruction).[4] We are now positioned at a critical moment in human history, at which we need to find this kind of creative equilibrium with the planet—especially between the extremes of unbridled consumption and population growth.

With the help of an earth literacy, we need to uncover, particularly in theology and religious studies, new modes of human–earth relations that incorporate the following three elements:

1. A worldview that understands that humans are secondary, the earth is primary (that is, the earth is primary scientist, healer, teacher, and technician).
2. A functional cosmology and an imaginative anthropólogy that reconceive the correspondences that are so pervasive in world religions of the relationship of the microcosm and the macrocosm (the human and the universe).
3. An environmental ethic that sees humans as subsystems of the earth system and thus aims to harmonize humans with the changes in the universe rather than to control humanity.

I would suggest that these three elements lead to a new kind of earth literacy for theology in light of what the sciences are telling us about the nature and complexity of the universe. This requires a new orientation to time and space in relation to the scientific story of the universe, which will have enduring implications for the study of theology and ethics, scriptures and ritual, and the history and comparison of religions. In short, it calls for a decentering and recentering of human processes in relation to earth processes.

First, reorienting with regard to *time:* A little over one hundred years after Darwin's discoveries, we are beginning to realize that we humans are part of a vast evolutionary process. It is still dawning on the human mind that we are products of developmental time, not simply seasonal time. This is a primary challenge to the most conservative Christian theologies, which have claimed their Genesis creation story, often without reference to the scientific theory of evolution. The pervasive character of evolution as developmental time also challenges liberal Christian theology, which has attempted to accommodate scientific thought by means of an anthropocentric vision of human progress. These issues need to be carefully and extensively rethought in all of their diverse implications for all of these schools of theology. Rather than collapsing theological thought into the science paradigm, we need to rethink the diverse religious implications of a contemporary functional cosmology. The work of Brian Swimme and Thomas Berry, as expressed in their volume *The Universe Story*,[5] is essential in this regard. This is the beginning of a *cosmological earth literacy*, which is already being discussed and critiqued in many circles.

Next, we are seeking reorientation with regard to *space*: Since the invention of the telescope and the microscope, we have been able to extend our vision outward to the galaxies and inward to the atom. New understandings of inner and outer space are emerging from these remarkable insights. Again, theology needs to be in dialogue with scientists on these issues.

We need, in particular, a literacy of place such as those Wendell Berry points toward in many of his essays. We need a grounding in a bioregion and all its diversity of life. Some of this implies simply a greater attention to natural history, biology, and botany. It is ironic that this kind of *bioregional earth literacy* has not been fostered. As the paleontologist Richard Leakey has observed:

> It is indeed a remarkable fact that we in this modern world, obsessed with measuring things, are so imprecise about the stuff of nature, to which we are intimately related and upon which we ultimately depend. We have a good estimate of how many stars there are in our galaxy, the Milky Way: some hundred billion. We know how many nucleotide bases constitute the human genetic blueprint: three billion. And we can calculate to within a few hours when a comet will collide with Jupiter, as it did at 4 PM (U.S. Eastern Daylight Time) on 16 July 1994.
>
> Yet we cannot put a secure number on current species diversity. It is not through lack of knowing how to obtain it but through lack of commitment. Governments have spent hundreds of millions of dollars in the systematic study of the stars but only a tiny fraction of that sum on a systematic study of nature here on Earth.[6]

The implications for new space–time understandings for theology and ethics are profoundly transformative. The explorations of a cosmological earth literacy with a bioregional earth literacy will be key to these fields in the future. In such a context, *theology* needs to see itself as not exclusively concerned with God–human or human–human relations, but also with human–earth relations.

Similarly, *ethics* needs to be expanded to embrace a love for species such as that described in E.O. Wilson's *Biophilia*.[7] This understanding of love of species will require scientists, such as Wilson, and theologians to reconsider ideas about religion and rights. Scientists need to understand that values of and meaningful relations with the natural world need not subvert the empirical powers of the human. Theologians need to consider that rights are not exclusive to the human sphere but that inherent sacred rights abide in other species as well. As Thomas Berry says, the basis of an environmental ethics is recognizing that the earth is "a communion of subjects not a collection of objects."

Similarly, scripture and ritual need rethinking: *Scripture* needs to be reconceived in the context of the scriptures of the universe. As Berry has observed, there are really four major scriptures:

- the cosmic scriptures of nature,
- the historical scriptures of human civilizations,
- the written scriptures of the world's great religions, and
- the scriptures of the human heart.

What is the role of scripture in this model? Have we ignored the scriptures of nature, and are we thus illiterate with regard to the natural world? How can we recover an appreciation for the scriptures of nature?

Ritual and liturgy also need to be envisioned as reflecting the cycles of nature and harmonizing humans with the earth. Paul Winter's *Earth Mass* and his solstice and equinox celebrations are important examples of this, and we seek more such life-giving rituals. We need to feel situated within the seasonal and cyclical rhythms of nature, especially in our highly industrialized and urbanized Western world.

Finally, *the history and comparison of religions* can contribute to earth literacy by examining ways in which different religions have conceived themselves in relation to nature. We need to explore the varied attitudes toward nature in the world's religions to reimagine ourselves in harmony with nature. To this end, a group of scholars organized ten conferences on religions of the world and ecology at Harvard's Center for the Study of World Religions.[8] The series had three concluding conferences in the fall of 1998, held at the American Academy of Arts and Sciences, the United Nations, and the American Museum of Natural History.[9]

Conclusion

Let me end these comments with a final note of cautious optimism. A work such as this is an important indicator of how far we have come on some of these issues—especially in the last five to ten years. I hope we will not talk at one another from two different sets of disciplines, but let us be open to reformulation and cooperative efforts of many kinds. And I ask us to think especially of what future generations are asking from us. I want to be able to share with my students signs that we have not abandoned intergenerational and interspecies concerns. Let us place ourselves amid the forces of entropy and energy to acknowledge our losses, to cherish our gains, and to hold forth the possibility of shifting our own destructive trajectory. With both profound grief and renewed hope I share the feelings of the poet Adrienne Rich, who wrote:

> My heart is moved by all I cannot save,
> So much has been destroyed.
> I have to cast my lot with those
> who, age after age, perversely,
> With no extraordinary power
> Reconstitute the world.

Notes

1. Daniel Maguire, *The Moral Core of Judaism and Christianity* (Philadelphia: Fortress Press, 1993), p. 13.
2. This term is used by Confucian scholar Tu Wei-ming in *Centrality and Commonality: An Essay on Confucian Religiousness* (Albany: SUNY Press, 1989).
3. Thomas Berry, *The Great Work* (New York: Belltower, 1999).
4. William McNeill, *The Human Condition: An Ecological and Historical View* (Princeton, N.J.: Princeton University Press, 1980).
5. Brian Swimme and Thomas Berry, *The Universe Story* (San Francisco: HarperSanFrancisco, 1992).
6. Richard Leakey, *The Sixth Extinction: Patterns of Life and the Future of Humankind* (New York: Doubleday, 1995).
7. E. O. Wilson, *Biophilia* (Cambridge, Mass.: Harvard University Press, 1984).
8. The papers from these conferences are being published by the Center and Harvard University Press.
9. See the Web site: http://divweb.harvard.edu/cswr/ecology.

Envisioning Equity in Consumption, Population, and Sustainability

Rodney L. Petersen*

Throughout this book we have tracked relationships between consumption, population, and sustainability through drawing into conversation the science and religion communities. Each of those three topics will shape policy debate in the twenty-first century. The correlation of patterns of consumption, population policy, and levels of acceptable sustainability is complex. While social theorists have worked at developing a calculus of the relationship, the variables are constantly shifting. Consumption is related to freedom. Population issues are related to our perception of the future. And each of these concepts carries emotional weight as they relate to our prevailing democratic philosophies and politics. Each draws us to reflect on the spheres of freedom that are open to individual decision making in comparison with the perceived necessity of central planning.[1]

The last quarter century has given an added edge to the variable of sustainability in discussions about consumption and population. This edge is the growing awareness among many individuals of ecological

*Rodney L. Petersen is executive director of the Boston Theological Institute. This essay grew out of work related to the conference "Consumption, Population & the Environment: Religion and Science Envision Equity for an Altered Creation," November 9–11, 1995, presented by the Boston Theological Institute together with the American Association for the Advancement of Science and funded through a grant from the Pew Charitable Trusts. A portion of this essay appeared as "Religion and the Quest for Equity in Consumption, Population, and Sustainability," *Bulletin of Science, Technology, and Society* vol. 19, no. 3 (June 1999), pp. 199–205.

degradation, whether at Chernobyl in the former Soviet Union, at Love Canal in the United States, or at sea with rogue vessels that carry the pollution of industrial waste. The nature and degree of this degradation affect all estimates of the sustainability or carrying capacity of the earth.[2] Societies throughout the world face environmental peril mixed with increasing social problems as the carrying capacity of the earth becomes increasingly taxed.

Religion and the Metaphors by Which We Live

This ecological challenge moves us to reconsider the metaphors by which we interpret our lives. How we use the earth and its resources or propagate our species in relation to all sentient life is related to our interpretation of ourselves and to the meaning we give to the world. Our worldview, while deeply personal and spiritual, is also derivative of the conceptual tools given to us by society. Religion, widely defined, is the formulation we give to a general order of existence. It might be said to be the answer that lies at the heart of every cultural question.[3] Religious perspectives are elaborated theologically, practiced socially, and applied ideologically as much to patterns of consumption and issues of population policy as to any other domain of our lives. Envisioning equity in matters of consumption, population, and sustainability draws us beyond the metaphor by which we live to the norms that we practice.

Many of our religions provide the premises for equity or social justice. As such, they lead us to the need for ecological justice or toward a shared and sustaining pattern of life. The path to eco-justice requires, at every step, not only attention to the motley and changing state of the world at a macro and local level, but also to the religious perspectives of all people. To envision equity in consumption, population, and sustainability is to call all people and their worldviews into dialogue together.[4] This indeed is what is happening in a variety of organizations such as the World Council of Churches, the Evangelical Environmental Network, the National Religious Partnership for the Environment, the North American Coalition on Religion and Ecology, and the Churches' Center for Theology and Public Policy.[5] Stated another way, ethics is related to epistemology, which is shaped by being. What we believe about ourselves along with our assumptions about human nature shape the way we live.[6]

The concern over issues of ecological degradation for the past quarter century has been paralleled by interest in the relationship of religion to issues of consumption and population in the context of a developing environmental ethic. This is particularly important insofar as religion helps to generate a worldview. One might chart a trajectory of contemporary concern that includes such world religious bodies as the World Council of Churches,[7] the Parliament of World Religions,[8] and particular denominations and academic departments of religion.[9] This mapping includes the scientific and technological community.[10] It includes intergovernmental agencies and nongovernmental organizations (NGOs) such as the World Bank.[11] In summary fashion, it includes the efforts of the Earth Charter Consultation Process to draft a shared vision of ethical values and practical guidelines essential to ecological security and sustainable living.[12]

Sensitivity to the Religious Perspectives of All People

Religious concern is not ancillary to this discussion but central and determining. It is important from a number of different directions. For example, one taxonomy of religions divides them by whether they exhibit monist or dualist tendencies. Monist religions tendencies emphasize the essential unity of all things, while those that are more dualist in orientation find division between, for example, this world and another world, the saved and the damned, etc. Having analyzed the universal religious concern through a variety of monist and dualist religions, Harold Coward concludes that aggressive, self-centered attitudes have typified human involvement with the environment worldwide. These attitudes are no longer acceptable, as the world's ecosystems are subject to increasing stress and potential collapse.[13] Of course, as Coward would agree, his picture should be nuanced. In this regard, Susan Power Bratton has surveyed seminal works in environmental and religious history from the 1960s and earlier.[14] Her reflections on mainstream Christian religion, various new theologies, Native American religions, and other forms of spirituality cause us to reconsider whether they are indeed consistently environmentally disruptive or if they are better characterized by "nature-friendliness."[15]

Within the Christian tradition, attitudes toward consumption and population are shaped by reading the Bible. This reading is interpreted through certain lenses.[16] For example, while the influence of Augustine of Hippo is often characterized by a "spiritualizing" of the Bible and hence a somewhat dualistic view of materiality,[17] Orthodox Christians have tended to approach nature with an ascetic and eucharistic attitude.[18] Among Protestants the interpretive spread is wide; Evangelicals, for example, are found in the array of positions from the "wise-use" movement on the extreme right to quite activist, communitarian, and even ascetic views represented among post-conservative Evangelicals.[19] Pro-natalist tendencies in Christianity have been tempered by this varying orientation.[20] Accordingly, patterns of consumption have varied widely with the degree of a "this-" or "other-worldly" orientation evident in different Christian communities.[21]

Additional lenses have been superimposed on this inherited Christian interpretation. Process theology, for example, has emerged as a way of discerning the evolving pattern of life and co-creative work of humanity with God.[22] Theologies of embodiment, which have tended to emphasize more immanental conceptions of God, have criticized the moderate dualism found in Augustine.[23] Feminist and womanist perspectives might also be cited.[24] Speaking out with one voice against an older theology that is conceived of in terms of the metaphor of male dominance,[25] feminist theologies develop positions that range from reference to Mother Earth and Gaia as poetic metaphors to new forms of dualism that find in Gaia a parallel to Isis, Astarte, and all of the "Great Mothers" of Antiquity.[26] While more "materialist" in orientation and affirming of political rights hitherto denied to women or other groups, such movements as these have often been characterized by a certain asceticism with respect to population and consumption.

Native peoples have been viewed as the first "bioregionalists" in matters of consumption,[27] sensitive to the web of life through a reciprocal system of bartering with the rest of sentient life for life,[28] and in matters of population conscious of not overtaxing the environment.[29] Diverse groups are a part of this discussion including the native experience in North America, the Celtic and Sammi peoples of Europe, the Maori of the South Pacific, tribal groups in Africa or South Asia, and indigenous peoples of Guatemala and Bolivia.[30] New age spirituality is often found in contemporary espousals of indigenous belief, particularly among "non-indigenous" persons.[31] This spirituality, together with that of deep ecology, might be set in an interpretive framework of increasing commitment to shaping patterns of consumption and population. The framework would include the following elements: (1) a "shallow" perspective that favors land conservation, stewardship, and the preservation of endangered species[32]; (2) an intermediate collection of approaches, including the land ethic of Aldo Leopold (an "eco-saint")[33] and the animal-liberation movement of Peter Singer,[34] which deny exclusive supremacy to human interests but still establish human responsibility for the ecosphere; (3) deep ecology, formulated by Arne Naess, William Devall, and George Sessions, calling for a new nonhuman ethic that would provide ecospheric equalitarianism and biodiversity for their own sakes[35]; and (4) deep-green ecology, which represents a deepening of deep ecology to a more critical and inductive style.[36]

There is a growing consensus that no single religious tradition or philosophical perspective has the solution to the issues associated with global sustainability.[37] Tu Wei-ming argues for a post-Enlightenment mentality that mobilizes the spiritual resources of the ethico-religious traditions (Greek philosophy, Judaism, Christianity), non-Western axial-age civilizations (Hinduism, Jainism, Buddhism, Confucianism, Taoism, Islam), and the spiritual resources of primal traditions.[38] This creates a wide field for discussion that is beginning to open up in society. It includes the question of how we live together in one world while holding different worldviews,[39] as important in matters of consumption and population as in settling regional violence that breaks out in different areas of the world. Such discussion is made more complex by the extent to which there is wide diversity within each religion on issues of environmental degradation and global sustainability.[40] All of this is quite apart from reflection on issues of cosmology, transcendence, and salvation, as these terms are conceived and defined differently among the great religions of the world.[41]

Nevertheless, the question of salvation and how it is understood within different religious traditions is important to debate over consumption and population policies respecting global sustainability. It is a question that carries us beyond a basic taxonomy of religion. If sustainability demands a global ethics, foundationalism in ethical reflection drives us to ask whether environmental needs are grounds enough for a sustained social policy. This is a difficult issue and will be raised here only for further consideration. The answer may be embraced by some from a reductionist or mechanistic orientation; a "green grace"— or nature as mediator of "God's" goodness and salvation—will be seen to be sufficient. For others, a deeper foundationalism—or a "red grace"—may be required, a grace that finds its grounding in a transcendent answer to personal and corporate identity. Such essentialism may be seen in traditional understandings of Semitic religion (Judaism, Christianity, and Islam). It may or may not be found in the perspectives of Buddhism or religious traditions that embrace the annihilation or absorption of being into Being. In a way that we have not seen before in human history, the question of the limits to the earth's carrying capacity is fostering a push toward a global environmental ethic and a deepening spirituality that asks of religion itself whether it is a product of culture or nature—or whether among the religions there is a spirit that transcends both culture and nature.[42]

The universality of religion seems evident from anthropology. However, this essentialism demands that we ask a further question: Is religion a subset of culture[43] or of nature?[44] Or does it stand in tension between the two, a natural impulse made manifest in culture?[45] It is not the purpose of this essay to resolve the question of the cultural nurture or inherent nature of religion. However, the question draws us to the issue of whether religion (or a religion) has its own unique epistemological voice. How we view this question is important not only to the ways in which we approach global sustainability, but also to our understanding of how dissent may arise surrounding social policy. We are learning too well how religion, often supportive of existing regimes, can also fuel social dissent and regional violence.

Religion's Contribution to Social Policy

Without settling the issues of duality/monism, the essential nature of religion, or its derivation in culture/nature, religion's contribution to social policy appears evident. First, the apparent crisis in sustainability is causing us to reflect more deeply upon relationships, an inherently religious concern. We are being pushed to ask about the nature of our relationship to community, in the widest sense as inclusive of all sentient life, and, by implication, to God or the "ground" of Being as it was defined by Paul Tillich, the theologian of culture.[46] Eco-ethicist Douglas John Hall puts this more particularly by asking us to consider whether we are above nature, in nature, or with nature. In his own analysis, he chooses the more synthetic, or third, position.[47] While most religions have been pro-natalist, some such as Buddhism and the aboriginal traditions have been more pointedly conscious of the limits defined by the web of life. Living with nature means that human fecundity cannot be at the expense of nature. This point is also inherent to the Semitic religions, if not always observed. Being conscious of relationships is two dimensional, relationships through time and relationships in contemporary space.

Second, how we live out these relationships becomes a matter of eco-justice. "Religion caring for creation" cannot mean an unbridled consumptive lifestyle. Rather, it is one lived in balance with the ecosystem in light of all human needs. If environmentalism is defined simply as wilderness preservation or the maintenance of biodiversity, it fails in terms of the relationships implied above. People are also natural. People are a part of nature. The concern for social and human health is as important as the general health, or sustainability, of the planet.[48] We are more aware today than ever before of just how deeply interrelated are issues of social and human health and of the environment around us.[49] The cleanliness of the water we drink, the conditions under which our food is grown or raised, and the integrity of our production and distribution systems have all been issues of increasing public concern. Liberation theology, once concerned largely with issues of political and economic justice, has been extended into the area of eco-justice by such well-known international ethicists as Leonardo Boff.[50] The impoverishment/environmental degradation cycle cannot be allowed to continue, for the sake of the earth as well as earth's people. If it does continue, religion will become an aspect of social dissent in a quest for eco-justice rather than supportive of established orders.

Our third concern takes us back to reflection on the essential nature of religion. For adherents of all religious traditions, and certainly for the Jewish and Christian traditions, the question of salvation in light of the environmental crisis is neither "green grace" nor "red grace" alone. Indeed, it is a question of grace to learn to live between what has been described as "the flood and the rainbow,"[51] between the crisis and the promise, which will require of us a new kind of courage in the twenty-first century. We who are on planet Earth are in something of an ark as we live between ecological disaster and the possibilities of a human community heretofore unimagined. In the face of the kinds of social reorientation that will be asked of us, social policy will wrestle with whether a grounding in environmental need is sufficient for a sustained social policy. Such policy will struggle with the issues common to all forms of utilitarianism in ethics. Cast in the light of sustainability, this includes the question of quantifying what is valuable in a society of growing social pressure. It includes the problem of knowledge in a day when we are recognizing the interdependence of all things. And it requires a sufficient comprehension of the extent of consequences in a society and history unparalleled in the human past. Religion's contribution to social policy, then, in relation to consumption, population, and sustainability is to help us to envision a policy for "earth community"[52] between the flood and the rainbow that attends to eco-justice for the whole community. This very struggle will draw us into deeper interfaith dialogue over the meaning of the metaphors through which we choose to live our lives.

The Rationality and Philosophy of Science

A carefully nuanced "foundationalism"—for lack of a more precise term, and not the foundationalism as critiqued rightly by deconstructionism and the post-structuralists—requires a certain degree of congruence between a religious and a scientific appreciation of the world.[53]

With such congruence, the path to ecological justice in consumption and population is made more sure. This is Audrey Chapman's contention as she argues for a faith, or religious vision, that is grounded in an informed science.[54] It is often clear enough today why science is important to religion. Less clear is the way in which scientism defines a worldview and replaces religion. For some, it is this world of dualistic (Cartesian) science, already wed to technology and market expansion, that is the problem behind the failure to deal with patterns of consumption and issues of population. For others, scientism defined as such has contributed to an "economic (European) religion" of the market, which distorts the real costs to the environment from the present practice of the exchange of goods and services.[55]

What we acknowledge as foundational will shape ethics. Whether such scientism has foisted upon the world a domineering anthropomorphism often blind to issues of ecojustice is not something to be resolved here. It can be said, however, that consumption has become divorced from moral philosophy and wedded to growth—defined as GNP and without regard for the wider issues of population and consumption that are thereby impacted. The recovery of a more holistic perspective is required. Whether religion as such, or a particular religion, can provide this wider vision for engaging ecologically singed issues may depend on whether a given tradition is seen as bearing signs of transcendence (symbolic instrumentalism) or as replete with symbols embedded in religious forms of life (linguistic pragmatism).

Although there were always voices questioning the relationship between science and a narrowing mechanistic positivism through the nineteenth century, European and Anglo-American societies grew to accept the division between facts and values that was increasingly practiced from the Enlightenment into the modern period—often for good reason. Writing with David Hume's epistemological skepticism in mind, Immanuel Kant's work and legacy put empirical knowledge on a firmer footing—but to the detriment of religious understanding, which was never satisfactory to Kant.[56] Although the "real" God escapes knowledge, as Kant defines God in his *Critique of Pure Reason* (1781),[57] the *idea* of God is valuable for speculative thought in at least three ways: (1) it helps to distinguish between appearances and things-in-themselves; (2) it suggests an explanation for the mystery of intuition; and (3) it promotes scientific inquiry through a confidence in the assumed intelligibility and unity of the world. Each of these three areas has now fallen subject to hermeneutical and cultural debate.[58] The wedding of science to positivism has not been a good match from an ecological perspective for either consumption or population. Even the criticism that science is contributing to a collapse of European values by fostering a spirit of negativity, detachment, and tentativeness appears tame today.[59]

Questions about the carrying capacity of the earth mandate a deeper conversation between the languages of science and religion. This conversation includes at least the following three observations. First, the language of "facticity" needs values. A coherent ethic for sustainability requires all the information that the sciences can muster. Yet, as Paul Ehrlich reminds us, above and beyond the technological and legislative changes that are mandated, the most important change that is required is a change in ourselves.[60] That such a dialogue is possible is the result of many startling discoveries about the nature of our world in the twentieth century and comes out of a different intellectual climate in the philosophy of science and the sociology of knowledge, which has developed since the Second World War.[61]

A second observation that might be made about such a conversation is that this change in intellectual climate has made for a more equal relationship between science and religion, one discerned in history rather than in the laws of determinacy. A critique of the limits of science has come not only from within the scientific community, but also from a variety of religious traditions.[62] Wolfhart Pannenberg finds the sciences drawn into a larger framework of intelligibility through the reflective discipline of theology.[63] He writes that increasing attention needs to be given to the relationship between natural laws and the contingency of individual events. Arguing in a way that parallels Michael Polanyi's idea of tacit knowledge, Pannenberg finds that scientific formulas, in whichever discipline they may be developed, ignore their contexts. This leads to the mistaken conclusion that the actual course of events is determined by the laws of nature, whereas contingency gets ignored. Nature, Pannenberg argues, ought to be understood as historical while natural laws should be seen as the uniformities abstracted from contingent (historical) events.[64] History—rather than determinacy—provides the "gate" for increased traffic between science and religion, notes theologian Ted Peters, adding that that is a space in which both theologians and practitioners of the new sciences are at home.[65]

Our third observation about the conversation between science and religion is that history implies a value placed on human activity. It also evokes the question of how a Creator—and perhaps humanity as well—participates in the management of nature. Pannenberg implies that the answer is found in the providential activity wherein God aims to accomplish God's tasks rather than a telos or entelechy; further, nature itself is to find its own fulfillment.[66] This idea relates to a point raised by the Australian biologist Charles Birch, who, drawing from Alfred North Whitehead, finds in process theology the conceptual tools for a theology of nature. However, governance may also imply resistance. This reminds us that in Christian theology creation is not an extension or emanation of God. It is an object of God's love, free to depart from or participate with God's purposes.[67] The arena for this drama is human activity in history. If history is the gate through which science and religion meet, we are drawn into an evolving narrative that includes conversation with all peoples of living faith over questions of consumption and population in the management of planet Earth.[68]

Envisioning Equity

Ethics have been worked out in recent times in relation to scientific knowledge (facts) and religious reflection (values).[69] These categories are two parts of a whole that frame the way we view and maintain sustainability within the environment.[70] The historical era into which we are moving challenges us to formulate a new ethic appropriate to the task before us.[71] The need for additional approaches to the kinds of issues raised in this volume on the part of society at large, government agencies, and churches and other voluntary agencies has become clear, particularly in relation to global economic structural concerns.[72] Earth's capacity to meet human demands for natural resources and, equally, its ability to absorb the waste produced by human activity are reaching their limits.[73] In addition to the kind of integration between science and religion called for by Chapman, the following strategies (among others) are required.

First, better long-range thinking is necessary to meet the environmental challenge.[74] This requires a cross-cultural scope because the nature of the problem is global. The search for a global ethic for environmental security and economic sustainability has implications for human rights in the two domains covered by this essay.[75] Any transnational thinking requires interreligious considerations.[76] This is particularly true for reflection on the Earth Charter, which seeks to identify the core values and principles that should guide global environmental conservation and sustainable development.[77] The debate entails derivative questions about human rights in the context of an emerging global politics, which demands that we move beyond the patterns of national interest that have dominated political thought at least since the Peace of Westphalia (1648).[78] It draws in definitions of development that affect economic well-being and reflect indigenous and other religious worldviews,[79] something that is beginning to happen among the churches with respect to the question of global climate change.[80] The debate entails the value of all sentient life. Such questions pick up the debate raised by "deep ecology" and "eco-philosophy,"[81] as well as the work of theologians like Andrew Linzey and Jay McDaniel.[82]

Second, further thought needs to be given to the meaning and use of technology. *Ecology* has to do with all living species, habitats, and ecosystems, while *environment* has to do with the human social, economic, and material context for life. Yet the terms are often used without discrimination or are collapsed into each other, interfering with our understanding of culture and nature. It is the premise of most theorists that the way we live in relation to these categories is both the locus of the problem and the solution to it. Politics and disputes about the meaning of environmental degradation become directly involved in the implications of economic action and technical deployment.[83] Our sense of the meaning and use of technology, raised elsewhere in this volume, is directly related to our understanding of the meaning of human activity in the world.

Our struggle to work ourselves out of a depressive determinism reaches into the soul of the environmental movement and the heart of the debate over sustainability. It is a discussion that can move in one of two directions. Our culture has the ability to solve its environmental problems. On the one hand, we might develop a neoliberal focus and try to solve the problem with the same tools that created it—with the market or central planning forces that gave rise to the crisis. In that case, we might expand the trading of pollution rights, allow for the development of wetlands, and sell off national lands to the highest bidder while maintaining some level of environmental law. On the other hand, we might focus on the problem of degradation itself and build into the prices of goods and services the full range of their costs (from raw resource to recyclable waste). Our worldview, however, might subordinate our sense of ourselves (culture) to a wider array of determining environmental factors (nature). In this case, we may allow that any human activity on behalf of the environment, the shaping of consumptive and population patterns, or the greater ecosystems of the earth is futile. The stronger interpretation of the Gaia hypothesis would allow that in the end the system will correct itself—if necessary, with the possible destruction of the human species. This variation of the nurture–nature controversy carries us back to the premodern period and debate over free will and individuality. How we think about ourselves is reflected in the meaning and use of technology.

The concept or metaphor that seeks to tie many of the religious traditions as they approach global sustainability is that of stewardship.

Embedded in Semitic tradition and developed variously, this image emphasizes the relational context in which humanity stands with respect to the rest of nature.[84] Through its emphasis on respect for all forms of life, it acknowledges a variety of worldviews and allows them to enter into conversation with one another.[85] Stewardship neither gives way to depressive determinism nor becomes overoptimistic about the spheres of human freedom open to us in the future.[86] It demands as full a knowledge of the natural world and its proper use as can be discerned.

Third, if stewardship defines our relational identity and draws us to our common task, further understanding is needed about the ways in which consumption and population relate to one another—particularly as both concepts touch so deeply upon our perceptions of individuality, freedom, and a sense of the future. For example, if it is argued that economic development for women is a means toward reducing levels of reproduction, does such "modernization" imply a higher level of consumption? Or again, the question of human reproduction has been an issue that is consigned to the private sphere in modern society; if it becomes a part of public policy, how might questions of human population become part of the "religion of the market"? And finally, in the movement toward equity in consumption and population within the parameters of global sustainability, what encouragement is there for those who have more to give to those who have less, particularly in light of human competitiveness? Such questions as these require a metaphor for human self-identity and behavior as is implied by the idea of stewardship. Apart from this metaphor, any idea of global harmony and justice as conceived by different religious traditions, whether a Jubilee Year, Sabbath Restitution, or Islamic Order, will never be realized.

Stewardship through Voluntary Associations and Churches

If the path ahead appears formidable, it is important to remember that decisions made hitherto have brought us to where we are and that different decisions can bring us to a different future.[87] The metaphor of stewardship in itself does not clearly define which path is best to take, but it can create a different mentality about and orientation to the problems we face. Stewardship can be lived out in a multitude of ways. However, it is best worked out in association with others as we work within the degree of freedom afforded to us. Associations may be communal, based upon natural and familial relationships. They have been increasingly voluntary in the evolution of modern society. Voluntary associations (independent associations, nongovernmental organizations, not-for-profit associations, public-regarding associations), which often are churches and, from a historical perspective, are most often birthed by churches, are central not only to any serious assessment of democratic public life but also to the demands of a life of stewardship. They are formed as people realize their own individual weakness in achieving positive social goals.[88]

Voluntary associations come into being and persist around issues of similarity of interest and of conscience. Insofar as issues of conscience draw people together, moral ends can be achieved through voluntary association. Churches—most pointedly—and many other voluntary societies are our corporate consciences as they struggle for identity between the individualism of contemporary society and the demands of the state.[89] As is being experienced around the world today in such fields as business ethics, in situations where there is no moral restraint on the part of individuals or corporations, the law must step in. With law comes a reduction of freedom, a diminishing of human autonomy. Failure to deal with our present patterns of consumption and population will bring legislation that will more narrowly codify the spheres of freedom open to people. And the failure to break the impoverishment/degradation cycle will have as its toll new levels of violence and repression.

It is an open question whether present governments alone can deal adequately with the kind of ethical agenda set before society, the need for long-range planning, a reassessment of the meaning and use of technology, and a deeper understanding of the relation between consumption and population. This inability is a part of the explanation for the dynamic behind an emerging world civic politics characterized by such organizations as Greenpeace, the World Wildlife Fund, and Friends of the Earth.[90] Churches and other voluntary societies are helping to give definition to a new united front through such organizations as the National Religious Partnership for the Environment. Churches, in the words of ethicist Larry Rasmussen, are learning that they must not be "*in* the way" of sustainability, but "*on* the way."[91] We maximize degrees of human freedom between the chaos of a degraded environment and the authority of law through the moral order advocated by voluntary associations. This space is, in fact, the place in which we work out our identity as people through the metaphor by which we understand our lives.[92]

Voluntary associations have a venerable history in North America. The historian of revivalism and religious movements Richard Lovelace has argued that what the environmental movement needs today is the kind of united front that was characteristic of the Evangelical united front in the nineteenth century, which did so much to eradicate black slavery in the Anglo-American world.[93] The combination of moral verve and spiritual energy, unrestrained by narrow denominational constraints, went far to change the social structures of the time. This kind of united front has been called for by Bruce Babbitt, United States secretary of the interior, in advocating a coalition of the religious and scientific communities with their characteristic emphases on values in pursuit of justice and facts.[94] The good news is that that is indeed happening.[95] Behind the rush toward spirituality at the onset of the twenty-first century is the struggle to find plausible relief to understand the dimensions of a situation that appears so bleak. Consumption is freedom, and population is our conception of the future. However, to fail to form such a united front is to run up the cost of a liveable environment beyond any imaginable inflationary level. An adequate understanding of stewardship can become our prevailing metaphor. A perception of ourselves that values human action can become our means. Long-range planning that understands the place of technology can be our method. Work in association with others becomes the movement.

Notes

1. In *The War Against Population* (San Francisco: Ignatius Press, 1988), Jacqueline Kasun summarizes the debate in favor of individual decision making within the context of Western liberal political thought and rule of market forces as against the need for social and economic administrative planning deemed necessary by Paul Ehrlich, author of *The Population Bomb*. See on the latter position, Titus Reid and David L. Lyon, eds., *Population Crisis: An Interdisciplinary Perspective* (Glenview, Ill.: Scott Foresman, 1972), and on the former position, Colin Clark, *Population Growth: The Advantages* (Santa Ana, Calif.: R. L. Sassone, 1972).

2. The World Commission on Environment and Development (WCED) defines *sustainability* as "development that meets the needs of the present without compromising the ability of future generations to meet their needs" (*Our Common Future* [New York: Oxford University Press, 1987], p. 43).

3. Ninian Smart, *Worldviews: Crosscultural Explorations of Human Beliefs* (New York: Scribner's, 1983), and more recently, *Dimensions of the Sacred* (Berkeley: University of California Press, 1996).

4. Currently, violence as related to religious motivation is more prevalent throughout the world than state-sponsored violence. The need for tolerance in society is as great or greater now than when John Locke wrote his *Letters on Tolerance* at the close of the seventeenth century. In a develop-

ing global society, sensitivity to religious diversity is as necessary as acknowledging theological difference. See John Locke, *Letters on Tolerance* (1689–92); cf. Patrick Collinson, "Religion and Human Rights: The Case of and for Protestantism," in Olwen Hufton, ed., *Historical Change and Human Rights: The Oxford Amnesty Lectures, 1994* (New York: Basic Books, 1995), pp. 21–54.

5. See the special issue of *Theology and Public Life,* vol. 8, nos., 1 & 2 (Summer and Winter 1996), entitled "The Ethics of Population, Consumption, and Environment: Essays and Cases."

6. The Worldwatch Institute continues to provide helpful yearly data in the *State of the World* reports, material that might be reflected upon through a number of different ethical matrices. See, e.g., Bernard Rosen, *Strategies of Ethics* (Boston: Houghton Mifflin, 1971). See James A. Nash, "Moral Values in Risk Decisions," in C. Richard Cothern, ed., *Handbook for Environmental Risk Decisionmaking* (Boca Raton, Fla.: Lewis Publishers, 1995), pp. 195–212.

7. Don Conroy and Rodney L. Petersen, eds., *Earth at Risk: Faith Communities in the Environmental Age* (Amherst, N.Y.: Humanity Books, forthcoming 2000).

8. G. O. Barney, *Global 2000 Revisited: What Shall We Do?* summary report on the critical issues of the twenty-first century, the Parliament of the World's Religions, Chicago, August 28–September 4, 1993.

9. See the series "Religions of the World and Ecology" facilitated by the Center for the Study of World Religions at Harvard University, Cambridge, Massachusetts, and edited by Mary Evelyn Tucker, Bucknell University.

10. For example, the *"Mission to Washington": The Joint Appeal by Religion and Science on the Environment,* which brought together religious leaders representing 330,000 U.S. congregations and some fifty scientists in the U.S. Senate on May 12, 1992, to find common cause in urging the U.S. government to better protect the global environment.

11. Ismail Serageldin and Richard Barrett, eds., *Ethics and Spiritual Values: Promoting Environmentally Sustainable Development,* Environmentally Sustainable Development Proceedings Series no. 12 (Washington, D.C.: World Bank, 1996).

12. The Earth Charter Drafting Committee, Steven Rockefeller, P.O. Box 648, Middlebury, Vt. 05753.

13. Harold Coward, ed., *Population, Consumption and the Environment: Religious and Secular Responses* (Albany: State University of New York Press, 1995), p. 12; see part 2.

14. Susan Power Bratton, "European Religion in America: The Undoing of the Environment?" in Conroy and Petersen, *Earth at Risk.*

15. The study of literature about the environment has spawned new academic disciplines, some of which work to draw together perspectives on the sciences and humanities and variously called ecocriticism, "green cultural studies," etc. See the journal *Interdisciplinary Studies in Literature and the Environment* and *The American Nature Writing Newsletter.* Related

debates ask whether such interdisciplinary work will mute the concerns of eco-feminism or other rights-based ethical debate.

16. H. Paul Santmire, *The Travail of Nature: The Ambiguous Ecological Promise of Christian Theology* (Minneapolis: Fortress, 1985). Santmire distinguishes two thrusts in the history of Christianity out of its Hebraic and Hellenistic contexts, a spiritual and an ecological motif in continuing tension with one another. See also Robert Booth Fowler, *The Greening of Protestant Thought* (Chapel Hill: University of North Carolina, 1995). In broad strokes the author traces the increasing influence of environmentalism on American Protestantism since the first Earth Day (1970).

17. Crawford Knox, *Changing Christian Paradigms and Their Implications for Modern Thought* (New York: E. J. Brill, 1993). The author focuses his work on the contributions of Augustine in shaping a particular way of reading the Bible in the West, arguing that pre-Augustinian and biblical understandings cohere more closely with the conceptions of modern science. The variety of perspectives among Roman Catholics is wider than what we encounter in the various manifestations of Orthodoxy ("The JPIC Process: A Catholic Contribution," *Ecumenical Review* 41, no. 4 [October 1989]: 591–602). The concept of the integrity of creation is found in Pope John Paul II's encyclical "Sollicitudo Rei Socialis" (1987), par. 26.

18. The ascetic nature of Orthodoxy suggests the need for a God-centered, prayerful self-discipline to curb our appetitive desires, particularly significant in light of population and consumptive patterns as they bear upon the carrying capacity of the earth. That Orthodoxy is awakening to its global ecological role was seen pointedly in 1989 when His Holiness Ecumenical Patriarch Dimitrios of Constantinople declared September 1, the first day of the Byzantine Ecclesiastical Year, as Environmental Protection Day (Gennadios Limouris, ed., *Justice, Peace and the Integrity of Creation: Insights from Orthodoxy* [Geneva: WCC Publications, 1990]).

19. An early Evangelical voice on behalf of ecological responsibility is that of Francis S. Schaeffer, *Pollution and the Death of Man: The Christian View of Ecology* (Wheaton, Ill.: Tyndale House Publishers, 1970); cf. more recently, J. Mark Thomas, guest editor, *Evangelicals and the Environment: Theological Foundations for Christian Environmental Stewardship,* a dedicated edition of the *Evangelical Review of Theology* 17, no. 2 (April 1993). See here for leading Evangelical voices on environmental stewardship and ecological concern. The Evangelical concern for linking issues of personal Christian discipleship with environmental ethics is seen in Vera C. Shaw, *Thorns in the Garden Planet: Meditations on the Creator's Care* (Nashville: Thomas Nelson, 1993). See Wesley Granberg-Michaelson, *A Worldly Spirituality: The Call to Take Care of the Earth* (San Francisco: Harper & Row, 1984).

20. Catherine Keller, "A Christian Response to the Population Apocalypse," in Coward, ed., *Population, Consumption and the Environment,* pp. 109–21.

21. One need only consider the impact of asceticism and its communal expression in monasticism.

22. The philosophy of Alfred North Whitehead has been of particular value in shaping the thinking of such process theologians as John Cobb and of biologist and philosopher Charles Birch.

23. Sallie McFague, *The Body of God: An Ecological Theology* (Minneapolis: Fortress Press, 1993).

24. Jane Cary Peck and Jeanne Gallo, "JPIC: A Critique from a Feminist Perspective," *Ecumenical Review* 41, no. 4 (October 1989): 573–81; Rosemary Radford Reuther, *Liberation Theology: Human Hope Confronts Christian History and American Power* (New York: Paulist Press, 1972), p. 115; and *Gaia and God: An Ecofeminist Theology of Earth Healing* (San Francisco: HarperCollins, 1992).

25. Contrary to the cautious attitude toward nature as a realm of immense mystery in medieval Europe, the seventeenth-century English philosopher Thomas Hobbes regarded nature as existing solely for human use: "She is no mystery, for she worketh by motion and geometry. . . . [We] can chart these motions. Feel then as if you lived in a world which can be measured, weighed and mastered and confront it with audacity" quoted in Basil Wiley, *The Seventeenth Century Background* (Garden City, N.Y.: Doubleday-Anchor, 1950), pp. 95–96.

26. Judith Plaskow and Carol P. Christ present helpful essays on understanding the nature of and motivation for goddess spirituality, in *Weaving the Visions: New Patterns in Feminist Spirituality* (San Francisco: Harper & Row, 1989), part 2. An alternative view is expressed by Loren Wilkinson, "Gaia Spirituality: A Christian Critique," in J. Mark Thomas, guest ed., *Evangelicals and the Environment: Theological Foundations for Christian Environmental Stewardship,* a special issue of *Evangelical Review of Theology* 17, no. 2 (April 1993): 176–89.

27. Jay B. McDaniel, *With Roots and Wings: Christianity in an Age of Ecology and Dialogue* (Maryknoll, N.Y.: Orbis, 1995), p. 68. Sensitive to Native American objections over the appropriation of their religious traditions, McDaniel has nevertheless been criticized for inappropriately importing native "spiritual lessons" into his own concerns.

28. Dennis McPherson and J. Douglas Rabb, *Indians from the Inside: A Study in Ethno-Metaphysics* (Thunder Bay, Ontario: Lakewood University Centre for Northern Studies, 1993), p. 90.

29. Daisy Sewid-Smith, "Aboriginal Spirituality, Population, and the Environment," in Harold Coward, ed., *Population, Consumption and the Environment,* pp. 63–71.

30. Thomas Berry with Thomas Clarke, *Befriending the Earth: A Theology of Reconciliation between Humans and the Earth* (Mystic, Conn.: Twenty-Third Publications, 1991); Jay B. McDaniel, *Earth, Sky, Gods and Mortals: Developing an Ecological Spirituality* (Mystic, Conn.: Twenty-Third Publications, 1990). This interest in the contribution of indigenous peoples

is particularly evident in programming at the World Council of Churches and in its study Theology of Life.

31. Vine Deloria, Jr., *God Is Red* (New York: Grosset and Dunlap, 1973). In Deloria's opinion, Christianity has forsaken nature, and an ecological era requires the spiritual resources of Native American religions for spiritual guidance. The recovery of Native spirituality is set more clearly in the context of the collapse of values in contemporary society in the second edition (Golden, Colo.: Fulcrum, 1992), pp. 52–53.

32. John Passmore, *Man's Responsibility for Nature* (London: Duckworth, 1974). The book was criticized for an anthropocentric framework and instrumentalist perspective on the natural world.

33. For example, see the land ethic of Aldo Leopold (*Sand County Almanac* [New York: Oxford University Press, 1949]) or the work of historian Lynn White (1967). Leopold claims that a biblically inspired Abrahamic ethic underlies our misuse of the land and that John Muir took Judeo-Christianity to task, arguing that narrow-minded religionists could not conceive of the idea that God cared for the rest of creation as well as humankind.

34. Peter Singer, *Animal Liberation: A New Ethics for Our Treatment of Animals* (New York: Avon, 1975).

35. Arne Naess, *Ecology, Community, and Life-Style* (Cambridge, England: Cambridge University Press, 1989) and William Devall and George Sessions, *Deep Ecology: Living as if Nature Mattered* (Salt Lake City, UT: G. M. Smith, 1985). Naess stresses life forms' self-realization, developed from Spinoza into a cosmology of biotic *Bildung*.

36. On forms of deep ecology and metaphysical naturalism, see George Sessions, "Shallow and Deep Ecology: A Review of the Philosophical Literature," in Robert C. Schultz, Jr., and Donald Hughes, eds., *Ecological Consciousness: Essays from the Earth Day X Colloquium, University of Denver, 21–24 April 1980* (Washington, D.C.: University Press of America, 1981), chapter 19.

37. This is the premise of the book edited by Mary Evelyn Tucker and John A. Grim, *Worldviews and Ecology: Religion, Philosophy, and the Environment* (Maryknoll, N.Y.: Orbis, 1994). See Harold Coward, ed., *Population, Consumption, and the Environment,* for the ways in which the major world religions view issues of population and resource consumption.

38. Tu Wei-ming, "Beyond the Enlightenment Mentality," in Tucker and Grim, eds., *Worldviews and Ecology,* pp. 19–29.

39. Samuel P. Huntington, *Clash of Civilizations and the Remaking of World Order* (New York: Simon & Schuster, 1996), pp. 47–48, 64–66.

40. George H. Williams explores the term *mercy* and its linguistic roots in a number of different cultures with the view of finding a cross-cultural basis for environmental ethics; see "Mercy in the Grounding of a Non-Elitist Ecological Ethic," in J. Sienkiwicz and James Betts, eds., *Festschrift in Honor of Charles Speel* (Monmouth, Ill.: Monmouth College, 1996).

41. The works of John Hick, Paul Knitter, and Gavin D'Costa mark out different positions in the debate about the unique nature of Christian salva-

tion. Special attention should be given to the ongoing discussion on gospel and culture facilitated by the World Council of Churches. See S. Wesley Ariarajah, *Gospel and Culture: An Ongoing Discussion within the Ecumenical Movement* (Geneva: WCC, 1994).

42. Burkert, *Creation of the Sacred: Tracks of Biology in Early Religions* (Cambridge, MA: Harvard University Press, 1996), pp. 1–33.

43. Clifford Geertz, *The Interpretation of Cultures* (New York: Basic Books, 1973). The perspective largely follows from the work of sociologist Emile Durkheim, *The Elementary Forms of Religious Life* (1918) and his idea of "collective representations." The continuing impact of the primacy given culture is seen in contemporary semiotics, structuralism, and poststructuralism.

44. See Burkert's discussion and general characteristics of religion, in *Creation of the Sacred*, pp. 5–8.

45. Unresolved here is whether religious symbols are instrumental, with the implication of transcendence, or pragmatic with respect to divinity (Pierce) or to religious forms of life (Wittgenstein and Derrida). See Robert C. Neville, *The Truth of Broken Symbols* (Albany: State University of New York Press, 1996), chs. 2–3.

46. Joseph Sittler, "Ecological Commitment as Theological Responsibility," *Zygon: Journal of Religion and Science* 5 (1970): 558. See Douglas Jon Hall's discussion, "The Ontology of Communion," in Hall, ed., *Imaging God: Dominion as Stewardship* (Grand Rapids: Eerdmans, 1986), pp. 113ff.

47. Douglas John Hall, *The Steward: A Biblical Symbol Come of Age* (Grand Rapids: Eerdmans, 1990), pp. 191–214. Sensitivity to the interconnectedness of things has been an attractive feature of some religions, e.g., see the Buddhist scholar Masao Abe, chapter in *Zen and Western Thought*, ed. William R. LaFleur (Honolulu: University of Hawaii Press, 1985).

48. A. J. McMichael, *Planetary Overload: Global Environmental Change and the Health of the Human Species* (Cambridge, England: Cambridge University Press, 1993).

49. Roger Gottlieb, *Forcing the Spring: The Transformation of the American Environmental Movement* (Washington, D.C.: Island Press, 1993).

50. Leonardo Boff, *Ecology and Liberation: A New Paradigm* (Maryknoll: Orbis Books, 1995). He writes, "The dominant trend of Christian reflection has not taken . . . creation to any profound level of consideration. For historical and institutional reasons, there has been much more consideration of redemption" (pp. 45, 47).

51. This is the title of the book compiled by D. Preman Niles, *Between the Flood and the Rainbow: Interpreting the Conciliar Process of Mutual Commitment (Covenant) to Justice, Peace and the Integrity of Creation* (Geneva: WCC, 1992). See also James A. Nash, *Loving Nature: Ecological Integrity and Christian Responsibility* (Nashville: Abingdon, 1991).

52. This is the title of a book by Larry Rasmussen, *Earth Community, Earth Ethics: Theology for the Third Millennium* (Maryknoll: Orbis Books, 1996).

53. The manner by which religion, or values as independent from the natural sciences, might be integrated is complex. From one point of view, it requires a worldview which is not that of scientism, although it may be scientific. The field of religious studies is one that is deeply divided between those for whom religion as an area of human engagement which is subject to natural and mechanistic explanation and others from whom such an unreservedly naturalistic approach begs the very premises of religious understanding. See J. van Baal, *Symbols for Communication: An Introduction to the Anthropological Study of Religion* (Assen: Van Gorcum 1971). Division in the field of anthropology between those who wish to maintain the aloofness of the social sciences from the natural sciences offers insight here. See Daniel Sperber, *Explaining Culture: A Naturalist Approach* (1996).

54. Audrey Chapman argues for four requirements for such integration. There needs to be 1) an adequate understanding of the fundamentals of modern science and its methodology; 2) an informed ethics developed in relation to contemporary biological and ecological sciences; 3) an understanding of specific issues; and 4) the delineation of new methodologies and approaches to integrating such analysis with religious perspectives. See *Rethinking Theology and Science: Six Models for the Current Dialogue*, ed. by Niels Henrik Gregersen and J. Wentzel Van Huyssteen (Grand Rapids: Wm. B. Eerdmans Publishing Co., 1998); also F. Herbert Bormann and Stephen R. Kellert, eds., *Ecology, Economics, Ethics: The Broken Circle* (New Haven: Yale University Press, 1991). The authors of this volume argue for the importance of a broad interdisciplinary approach that draws together the concepts raised up in the title to their book. Short-term environmental policy fails to take into account the economic or moral burden that is being placed upon future generations through the present depletion of resources.

55. A. Rodney Dobell, "Environmental Degradation and the Religion of the Market," in *Coward, Population, Consumption, and the Environment*, pp. 229–250.

56. Following the writing of the *Critique of Pure Reason* (1781), Kant wrote his *Critique of Practical Reason* (1788) in which he discerned a "felt" need for religion which results from a moral law. This moral functionalism became the basis for a moral theology of use in the world of values if not in that of facts. In *Religion Within the Limits of Reason Alone* (1793), Kant admits no supernatural revelation but equates Christian theology with the religion of practical reason. A modern restatement of this might be seen in the systematic theology of Gordon Kaufman, *God the Problem* (Cambridge: Harvard University Press, 1972), criticized for its "residual Cartesianism." More recently Kaufman has drawn his attention to constructive reflection around the area of eco-theology.

57. On the basis of knowledge, defined by Kant in the a priori categories of understanding (reason) together with empirical data (experience), meta-

physics is shown not to be a genuine science and arguments for God's existence speculative.

58. Concern for the rationality of science in light of its current detractors can be seen in the Conference "The Flight From Science and Reason," sponsored by The New York Academy of Sciences, 31 May–2 June 1995. See n. 20 below.

59. The three points are raised by Nels F. S. Ferré in the context of his article, "The Immorality of Science," *Religion in Life* 10, no. 1 (Winter 1941): 31–40. This critique is applied specifically to the emergence of scientific positivism, as well as to other concerns. Similar concerns with respect to technology are developed by the sociologist Jacques Ellul: "Technique is the translation into action of man's concern to master things by means of reason, to account for what is subconscious, make quantitative what is qualitative, make clear and precise the outlines of nature, take hold of chaos and put order into it" (p. 43). This mentality has become, for Ellul, the reigning mythology of our epoch. See his study, *The Technological Society* (New York: Alfred A. Knopf, 1964).

60. Paul Ehrlich, *The Machinery of Nature* (New York: Simon & Schuster, 1986), p. 17. Ehrlich writes: "I am convinced that such a quasi-religious movement, one concerned with the need to change the values that now govern much off human activity, is essential to the persistence of our civilization."

61. Philosophers of meaning such as Wilhelm Dilthey, Hans-Georg Gadamer, and Jürgan Habermas have underscored the notion that all experience of meaning participates in with widest context of meaning. Pannenberg develops this point by arguing that God is the all-determining reality and is the hypothesis that explains most adequately the whole experience of reality.

62. Lesslie Newbigin, *Foolishness to the Greeks: The Gospel and Western Culture* (Geneva: World Council of Churches, 1986), pp. 65–94. An understanding of critical realism as a place where a philosophy of science and theology might meet is given by W. van Huysteen in *Theology and the Justification of Faith* (Grand Rapids: Eerdmans, 1989), ch. 9; and in Michael Banner, *The Justification of Science and the Rationality of Religious Belief* (Oxford: Clarendon Press, 1990).

63. Wolfhart Pannenberg, *Toward a Theology of Nature: Essays on Science and Faith*, ed. by Ted Peters (Louisville: Westminster/John Knox Press, 1993). For further examples, see the work dedicated to the Society of Ordained Scientists by biochemist Arthur Peacocke, *Theology for a Scientific Age: Being and Becoming—Natural and Divine* (Oxford: Basil Blackwell, 1990). In making his case for theology as a science in dialogue with the natural sciences, Pannenberg offers a careful analysis of the terms *naturwissenschaften* and *geisteswissenschaften* in *Theology and the Philosophy of Science*, trans. F. McDonagh (Philadelphia: Westminster Press, 1976, p. 72; more fully in his *Systematic Theology* (Grand Rapids: Eerd-

mans, 1991). See also the early work of David Tracy, *Blessed Rage for Order* (New York: Seabury, 1975); and Bernard Lonergan, *Insight: A Study of Human Understanding* (New York: Philosophical Society, 1958).

64. See Pannenberg, "God and Nature," trans. by Wilhelm C. Linss, in *Toward a Theology of Nature*, pp. 50–71.

65. Peters writes, "To the theologian, the enduring forms of nature right along with single events appear as the contingent product of the activity of a free God." See his introductory essay in *Toward of Theology of Nature*, p. 10.

66. Ted Peters, whom I am following here, contrasts this with the medieval (Thomas Aquinas) purpose of the visio Dei whereby God in God's self is goal or scholastic Protestantism that finds praise of God as the chief end of creation. Both ideas proximate concepts of divine narcissism in Peters's view, in *Toward a Theology of Nature*, p. 11.

67. Many different ways have been developed to express this. Perhaps the most graphic is the idea of "the Omega Point," as developed by Teilhard de Chardin, *Hymn of the Universe* (New York: Harper & Row, 1965) and *The Phenomenon of Man* (New York: Harper and Row). Other models of God's interaction with the world are presented in Peacocke, *Theology for a Scientific Age*, pp. 135–183; for preaching, see Thomas F. Torrance, *Preaching Christ Today: The Gospel and Scientific Thinking* (Grand Rapids: Eerdmans, 1994), pp. 41–71.

68. William Clark, "Managing Planet Earth," in *Managing Planet Earth: Readings from Scientific American Magazine* (New York: W. H. Freeman and Co., 1989). Clark argues that the basic environmental issues are ethical and not scientific.

69. On the role of ethics with respect to sustainability, see J. Ronald Engel and Joan Gibb Engel, eds., *Ethics of Environment & Development: Global Challenge, International Response* (Tucson: University of Arizona, 1993); note the series of reports issued by the Institute for Philosophy and Public Policy, School of Public Affairs, University of Maryland, College Park, Maryland, e.g., *The Ethics of Consumption* (David A. Crocker and Toby Linden, eds., Lanham, MD: Rowman & Littlefield, 1998). See the debate over Act theories of ethical reflection, which deny the necessity of rule theories, whether teleological or deontological in Bernard Rosen, *Strategies of Ethics* (Boston: Houghton Mifflin, 1978)

70. Anne Whyte, "The Human Context," *Coward, Population, Consumption, and the Environment*, 52.

71. Ethical reflection in the churches of the twenty-first century must move more profoundly beyond the three tendencies identified by Paul Albrecht over the past seventy-five years: 1) an ethic of agapé going back to the origins of the *Life and Work Movement* (Stockholm, 1925); 2) a participatory and populist ethic with an overriding concern for those oppressed through racial, religious, or sexual identity; and 3) liberation theology, guided by a Marxist "science" in matters of praxis with a preference for the poor. See Paul Albrecht, "From Oxford to Nairobi: Lessons From Fifty

Years of Ecumenical Work for Economic and Social Justice," *The Ecumenical Review* 40 (April 1988).

72. For example, Larry Rasmussen, *Earth Community, Earth Ethics* (Maryknoll: Orbis Books, 1996); see Ans van der Bent, *Commitment to God's World* (Geneva: WCC Publications, 1995); see David G. Hallman, *Ecotheology: Voices from South and North* (Maryknoll: Orbis, 1994).

73. Anne Whyte, "The Human Context," Coward, ed., *Population, Consumption, and the Environment*, p. 48.

74. World Commission on Environment and Development, *Brundtland Report* (1987), *Our Common Future* (New York: Oxford University Press, 1987). The Report notes the rapid deterioration of the global environment as threatening human life on earth. It seeks to delineate approximate and possible ways to deal with environmental issues. It stands for 1) meeting the needs of the present without compromising the ability of future generations to meet their needs; 2) creating a sustainable situation for all countries; and 3) a concern for equality within and between generations, not just physical sustainability. This was the third in a series of UN reports, following the Brandt report North–South (1980) with its sequel *Common Crisis* (1983) and the Palme report *Common Security* (1985). It deals with major problems like the greenhouse effect, deforestation, soil loss, the debt crisis, the global commons, and the explosion of cities.

75. Hans Küng, *Global Responsibility: In Search of a New World Ethic* (New York: Continuum, 1993).

76. John Leslie insists that environmental issues are the national security issues of the twenty-first century, in *The End of the World: The Science and Ethics of Human Extinction* (New York: Routledge, 1996); cf. Samuel Huntington, *The Clash of Civilizations and the Remaking of World Order* (New York: Simon & Schuster, 1996), pp. 130-35.

77. "The Earth Charter: A Joint Initiative of the Earth Council and Green Cross International," a report prepared by the Earth Council and Green Cross International in connection with the Earth Charter Workshop, The Peace Palace, The Hague, May 31, 1995; and see the document in progress, *Principles of Environmental Conservation and Sustainable Development: Summary and Survey*, prepared for the Earth Charter Project by Steven Rockefeller (Earth Charter Project, The Earth Council, San Jose, Costa Rica, April 1996).

78. Andrew Harrell and Benedict Kingsbury, eds., *The International Politics of the Environment* (Oxford, 1992); and Paul Wapner, *Environmental Activism and World Civic Politics* (Albany: State University of New York Press, 1996).

79. See the papers from the conference "A Religious and Moral Challenge. Environmental Justice." *A Briefing Sponsored by The National Religious Partnership for the Environment, January 9 and 10, 1995. See Sustainable Growth—A Contradiction in Terms? Report of the Visser 't Hooft Memorial Consultation* (Chateau de Bossey: The Ecumenical Institute, 1993).

80. *Overview of the World Council of Churches' Programme on Climate Change* (Geneva: WCC, 1995).

81. Richard Sylvan and David Bennett, *The Greening of Ethics: From Anthropomorphism to Deep-Green Theory* (Tucson: The University of Arizona Press, 1994). Henryk Skolimowski develops an ecological humanism as an alternative to industrial society that sees 1) stewardship as a prevailing metaphor for human activity in the future, 2) the world to be best conceived of as a sanctuary in a religious sense, and 3) knowledge to be defined as the intermediary between us and the creative forces of evolution. See *Eco-Philosophy: Designing New Tactics for Living* (London: Marion Boyars, 1981), p. 54.

82. See, for example, Andrew Linzey's *Christianity and the Rights of Animals* (New York: Crossroads, 1987) and idem, *Animal Theology* (Urbana: University of Illinois Press, 1995). McDaniel's book *On God and Pelicans* (Philadelphia: John Knox Press, 1989) advocates a life-centered ethics, which gains from the insights of process thought, the animal rights movement, and feminist theology. See also idem, *With Roots and Wings*, 1995.

83. I am following Goldblatt, *Social Theory and the Environment* (Oxford: Oxford University Press, 1996).

84. See Part II, a survey of different religious traditions, in *Coward, Population, Consumption, and the Environment,* pp. 63–194. In developing his ideas on stewardship, Douglas John Hall contends that a form of Christian humanism that transcends the theocentric as well as the older liberal perspective (faulted for its assumptions about humanity and history and failure to provide an acceptable theology of nature) must be pioneered. See his *The Steward: A Biblical Symbol Come of Age* (Grand Rapids: Eerdmans, 1994 ed.), pp. 103–21. In order to enlarge our vision of the full ramifications for stewardship, he cites the following five principles: globalization, communalization, ecologization, politicization, and futurization (pp. 122–54). See the modified or "weak" anthropocentrism of Holmes Rolston III, *Environmental Ethics: Duties to and Values in the Natural World* (Philadelphia: Temple University Press, 1988).

85. See the discussion in Section Two about "Living with Nature." On the problem of the disconnection between ecological belief and behavior, see Walt Grazer, "The Vision that Heals," *Green Cross,* vol. 2, no. 4 (Fall 1996): 16–21.

86. In referring to the potential latent in the metaphor of stewardship, Douglas John Hall writes that as Martin Luther's vision of a justifying grace enabled people to overcome a sense of medieval guilt and thereby find the courage to live, so too "the sense of being stewards of earth and of life itself could provide a generation of world-weary and apathetic survivors some feeling of purpose." See Hall, *The Steward,* p. 7. See the author's comments about the history of the book, first published in 1982, with reference to the emergence of the "Justice, Peace, and the Integrity of Creation" process in the World Council of Churches (xii). A helpful study

guide is available to Hall's understanding of stewardship by J. Phillips Williams, *A Study Guide for Douglas John Hall's The Steward: A Biblical Symbol Come of Age* (New York: Friendship Press, 1985). Other metaphors include those of "frugality " and "mercy." On the former, see James A. Nash, "Toward the Revival and Reform of the Subversive Virtue: Frugality," earlier in this book; and the latter, George H. Williams, "Mercy in the Grounding of a Non-Elitist Ecological Ethic," in *Festschrift in Honor of Charles Speel,* ed., Thomas J. Sienkewicz and James Betts (Monmouth, IL: Monmouth College, 1996).

87. Jael M. Silliman, "Ethics, Family Planning, Status of Women, and the Environment," in Coward, ed., *Consumption, Population, and the Environment,* p. 255.

88. On the nature of voluntary associations, see the important volume edited by J. Ronald Engel, *Voluntary Associations: Socio-cultural Analyses and Theological Interpretation: James Luther Adams* (Chicago: Exploration Press, 1986).

89. By analogy, the Reformed theologian John Calvin maintained in his interpretation of I Samuel 8:4-18 that Israel was better ruled by conscience and the association of judges than under a king with a king's authority and rule. John Calvin, *Institutes of the Christian Religion,* ed. John T. McNeill, trans. Ford Lewis Battles, vol. 1 (Philadelphia: Westminster Press, 1960), 4.xx.1–26. For contemporary issues that pertain, see Robert Bellah, "How to Understand the Church in an Individualistic Society," in Petersen, ed., *Christianity and Civil Society* (Maryknoll: Orbis Books, 1995), pp. 1–14.

90. Paul Wapner, *Environmental Activism and World Civic Politics* (Albany: State University of New York Press, 1996), pp. 152–164. A brief history of the environment movement is presented by John Young, *Sustaining the Earth* (Cambridge, MA: Harvard University Press, 1990).

91. Larry L. Rasmussen, *Earth Community, Earth Ethics,* p. 349.

92. Charles A. Taylor, *The Sources of the Self: The Making of the Modern Identity* (Cambridge, MA: Harvard University Press, 1989). An illustration of the way in which Christian values are worked out in relation to human population in the context of environmental ethics is in Susan Power Bratton, *Six Billion and More: Human Population Regulation and Christian Ethics* (Philadelphia: Westminster/John Knox Press, 1992).

93. Richard F. Lovelace, *Dynamics of Spiritual Life: An Evangelical Theology of Renewal* (Downers Grove, IL: Inter-Varsity press, 1979), pp. 364–374.

94. See excerpts from Babbitt's remarks at the conference, *Consumption, Population & the Environment: Religion and Science Envision Equity for an Altered Creation,* sponsored by the Boston Theological Institute and the American Association for the Advancement of Science, "Living in Nature: Religion & Science in dialogue on the Environment" (Kansas City: Sheed & Ward, 1997).

95. For example, the National Religious Partnership for the Environment was brought together for just such a purpose in 1992 in response to an "Open Letter to the Religious Community" (January 1990), written by thirty-four

prominent scientists, and the subsequent "Joint Appeal" (1991) by persons representing the religious, scientific, and public policy communities. It represents constituencies defined by the U. S. Catholic Conference, the National Council of Churches of Christ, the Coalition on the Environment and Jewish Life, and the Evangelical Environmental Network. For a broader look at the way in which NGOs are emerging to define the social and political "space" between the family and the state in service of issues of ecology as well as immigration and relief, seeking solutioins to regional conflict, and other global problems, see John Tirman, "Forces of Civility. The NGO Revolution and the Search for Peace," *Boston Review* (December/January, 1998–99).

About the Contributors

Abdul Cader Asmal, Chairman of Communications, Islamic Council of New England

Mohammed Asmal, M.D./Ph.D. Candidate, Columbia University

Bruce Babbitt, Secretary, United States Department of the Interior

Ian G. Barbour, Professor of Religion and Bean Professor of Science, Technology, and Society Emeritus, Carleton College

Susan Power Bratton, Lindaman Chair of Science, Technology, and Culture, Whitworth College

Michael Brower, President, Brower & Company

David M. Byers, Executive Director, Committee on Home Missions; Committee on Science and Values

Audrey R. Chapman, Director, Program of Dialogue Between Science and Religion; Director, Science and Human Rights Program, American Association for the Advancement of Science

Emmanuel Clapsis, Associate Professor of Dogmatics, Holy Cross Greek Orthodox School of Theology

Joel E. Cohen, Professor of Populations, The Rockefeller University and Columbia University

David A. Crocker, Senior Research Scholar, Institute for Philosophy and Public Policy and School of Public Affairs, University of Maryland

Neva R. Goodwin, Co-Director, Global Development And Environment Institute, Tufts University

Warren Leon, Deputy Director for Programs, Union of Concerned Scientists

James Martin-Schramm, Assistant Professor of Religion, Luther College

James A. Nash, retired Executive Director, The Churches' Center for Theology and Public Policy

Rodney L. Petersen, Executive Director, Boston Theological Institute

William E. Rees, Professor and Director, School of Community and Regional Planning, The University of British Columbia

Michal Fox Smart, Jewish Environmental Educator

Barbara Smith-Moran, S.O.Sc., Director, Faith and Science Exchange, Boston Theological Institute

Mary Evelyn Tucker, Professor of Religion, Bucknell University

Index